W9-CLD-981

COOPERATIVE RESEARCH
IN THE NATIONAL MARINE FISHERIES SERVICE

Committee on Cooperative Research
in the National Marine Fisheries Service

Ocean Studies Board

Division on Earth and Life Studies

NATIONAL RESEARCH COUNCIL
OF THE NATIONAL ACADEMIES

THE NATIONAL ACADEMIES PRESS
Washington, D.C.
www.nap.edu

THE NATIONAL ACADEMIES PRESS 500 Fifth Street, N.W. Washington, DC 20001

This paper is funded in part by a contract from the National Oceanic and Atmospheric Administration. The views expressed herein are those of the authors and do not necessarily reflect the views of NOAA or any of its subagencies.

International Standard Book Number 0-309-09074-1 (Book)
International Standard Book Number 0-309-52747-3 (PDF)
Library of Congress Catalog Number 2003114991

Additional copies of this report are available from the National Academies Press, 500 Fifth Street, N.W., Lockbox 285, Washington, DC 20055; (800) 624-6242 or (202) 334-3313 (in the Washington metropolitan area); Internet, http://www.nap.edu

THE NATIONAL ACADEMIES
Advisers to the Nation on Science, Engineering, and Medicine

The **National Academy of Sciences** is a private, nonprofit, self-perpetuating society of distinguished scholars engaged in scientific and engineering research, dedicated to the furtherance of science and technology and to their use for the general welfare. Upon the authority of the charter granted to it by the Congress in 1863, the Academy has a mandate that requires it to advise the federal government on scientific and technical matters. Dr. Bruce M. Alberts is president of the National Academy of Sciences.

The **National Academy of Engineering** was established in 1964, under the charter of the National Academy of Sciences, as a parallel organization of outstanding engineers. It is autonomous in its administration and in the selection of its members, sharing with the National Academy of Sciences the responsibility for advising the federal government. The National Academy of Engineering also sponsors engineering programs aimed at meeting national needs, encourages education and research, and recognizes the superior achievements of engineers. Dr. Wm. A. Wulf is president of the National Academy of Engineering.

The **Institute of Medicine** was established in 1970 by the National Academy of Sciences to secure the services of eminent members of appropriate professions in the examination of policy matters pertaining to the health of the public. The Institute acts under the responsibility given to the National Academy of Sciences by its congressional charter to be an adviser to the federal government and, upon its own initiative, to identify issues of medical care, research, and education. Dr. Harvey V. Fineberg is president of the Institute of Medicine.

The **National Research Council** was organized by the National Academy of Sciences in 1916 to associate the broad community of science and technology with the Academy's purposes of furthering knowledge and advising the federal government. Functioning in accordance with general policies determined by the Academy, the Council has become the principal operating agency of both the National Academy of Sciences and the National Academy of Engineering in providing services to the government, the public, and the scientific and engineering communities. The Council is administered jointly by both Academies and the Institute of Medicine. Dr. Bruce M. Alberts and Dr. Wm. A. Wulf are chair and vice chair, respectively, of the National Research Council.

www.national-academies.org

Preface

In recent years there has been growing interest in having fisheries stakeholders involved in various aspects of fisheries data collection and experimentation. This activity is generally known as cooperative research and may take many forms, including gear technology studies, bycatch avoidance studies, and surveys. While the process of cooperative research is not new at all, the current interest in cooperative research and the growing frequency of direct budgetary allocation for cooperative research prompted the National Marine Fisheries Service (now National Oceanic and Amospheric Administration Fisheries) to commission this study.

The committee consisted of individuals from a broad range of disciplines and perspectives, including academic scientists, a Canadian government scientist, a non-governmental organization scientist, an international commission scientist, commercial fishermen, a fisheries management council member, and private consultants. All of us had some experience with various forms of cooperative research, and we were all interested in exploring what circumstances were appropriate for cooperative research and what it would take to make cooperative research the most effective.

The committee met three times (Boston, Seattle, St. Petersburg), and at each location the committee heard testimony from a range of participants in a variety of cooperative research projects. The committee listened to successes and failures and heard from individuals who were very supportive of cooperative research and people who had had unhappy experiences. While the examples the committee heard varied from region to region, there was a general consistency throughout the country about what worked and did not work as cooperative research.

The committee would like to thank those individuals who provided testimony to the committee. I would like to thank the committee members for their time and effort and the National Research Council staff, Terry Schaefer and Denise Greene, for their work.

Ray Hilborn
Chair

Acknowledgments

This report was greatly enhanced by the participants of the multiple information-gathering activities held as part of this study. The committee would first like to acknowledge the efforts of those who gave presentations at the committee meetings. These talks helped set the stage for fruitful discussions in the closed sessions that followed.

Daniel Cohen, Atlantic Cape Fisheries; Eric Powell, Rutgers University; Dave Wallace, Wallace & Associates; Jim Weinberg, Northeast Fisheries Science Center; Hans Davidsen, commercial fisherman; Bill Dupaul, Virginia Institute of Marine Science; Paul Rago, National Marine Fisheries Service; Ron Smalowitz, consultant; Chris Glass, Manomet Center for Conservation Sciences; Troy Hartley, Northeast Consortium; Earl Meredith, National Marine Fisheries Service; Craig Pendelton, Northwest Atlantic Marine Alliance; Don Perkins, Gulf of Maine Aquarium; Laura Taylor-Singer, Gulf of Maine Aquarium; Robert Tetrault, commercial fisherman; Wendy Gabriel, National Marine Fisheries Service; David Goethel, commercial fisherman; Harry Mears, National Marine Fisheries Service; Anthony Chatwin, Conservation Law Foundation; Ilene Kaplan, Union College; Bruce Turris, Canadian Sablefish Association & Canadian Groundfish Research and Conservation Society; Gary Stauffer, Alaska Fisheries Science Center; Arne Fuglvog, commercial fisherman; John Gauvin, Groundfish Forum; Jim McManus, Trident Seafoods; Gary Painter, commercial fisherman; Elizabeth Clarke, Northwest Fisheries Science

Center; Jennifer Bloesser, Pacific Marine Conservation Council; Terry Thompson, commercial fisherman; Danny Parker, commercial fisherman; Vidar Wespestad, Resource Analysts, International; Ralph Brown, commercial fisherman; Bruce Leaman, International Pacific Halibut Commission; Julia Parrish, University of Washington; Craig Rose, Alaska Fisheries Science Center; Chris Lunsford, Alaska Fisheries Science Center; Paul Starr, New Zealand Seafood Industry Council; Martin Hall, Inter-American Tropical Tuna Commission; Gerry Scott, Southeast Fisheries Science Center; Benny Gallaway, LGL Ecological Research Associates, Inc.; Robert Hueter, Mote Marine Laboratory; Geof Lane, Gulf and South Atlantic Fisheries Foundation; Scott Nichols, Southeast Fisheries Science Center; Pascagoula Laboratory; Sal Versaggi, Versaggi Shrimp Company; Judy Jamison, Gulf and South Atlantic Fisheries Foundation; Kerry O'Malley, South Atlantic Fishery Management Council; Steve Rebach, North Carolina Sea Grant; Ellie Roche, National Marine Fisheries Service, Southeast Regional Office; Eric Sander, Florida Marine Research Institute; Karen Burns, Mote Marine Laboratory; Greg Didomenico, Monroe County Fisherman, Inc.; Bobby Spaeth, Madeira Beach Seafood; John Hunter, National Marine Fisheries Service, La Jolla Laboratory; David Demer, National Marine Fisheries Service, La Jolla Laboratory; Gerald DiNardo, National Marine Fisheries Service, Honolulu Laboratory; David Hamm, National Marine Fisheries Service, Honolulu Laboratory.

This report has been reviewed in draft form by individuals chosen for their diverse perspectives and technical expertise, in accordance with procedures approved by the National Research Council's Report Review Committee. The purpose of this independent review is to provide candid and critical comments that will assist the institution in making the published report as sound as possible and to ensure that the report meets institutional standards for objectivity, evidence, and responsiveness to the study charge. The review comments and draft manuscript remain confidential to protect the integrity of the deliberative process. We wish to thank the following individuals for their participation in the review of this report:

Vaughn Anthony, retired National Marine Fisheries Service assessment
    biologist, Boothbay, Maine
Jennifer Bloesser, Pacific Marine Conservation Council, Arcata,
    California

John Gauvin, Groundfish Forum, Seattle, Washington
Christopher Glass, Manomet Center for Conservation Sciences, Manomet, Massachusetts
Judy Jamison, Gulf & South Atlantic Fisheries Foundation, Tampa, Florida
Seth Macinko, University of Rhode Island, Kingston, Rhode Island
Bonnie McCay, Rutgers University, Cook College, New Brunswick, New Jersey
Andrew Solow, Woods Hole Oceanographic Institution, Woods Hole, Massachusetts
Robert Schoning, retired, former director, National Marine Fisheries Service, Corvallis, Oregon
Richard Young, commercial fisherman, Crescent City, California

Although the reviewers listed above provided many constructive comments and suggestions, they were not asked to endorse the conclusions or recommendations nor did they see the final draft of the report before its release. The review of this report was overseen by Robert Frosch, Harvard University, Cambridge, Massachusetts. Appointed by the National Research Council, he was responsible for making certain that an independent examination of this report was carried out in accordance with institutional procedures and that all review comments were carefully considered. Responsibility for the final content of this report rests entirely with the authoring committee and the institution.

Contents

xi

8 WHAT WORKS AND DOESN'T WORK   **95**
Reasons for Success, 95
Reasons for Failure, 106
Making Cooperative Research Work, 108

9 FINDINGS AND RECOMMENDATIONS   **111**
Evaluating the Benefits of Cooperative Research, 111
Allocation of Funding, 112
Legal and Administrative Issues, 114
Communication, 117

REFERENCES   **119**

APPENDIXES

Executive Summary

In the United States and around the world, the dominant method for obtaining information for the management of fisheries has been through centralized, government-staffed research programs. There is a growing trend to involve other parties in fisheries research activities under the general umbrella of "cooperative research." These parties include commercial and recreational fishermen, fishing industry groups, nongovernmental organizations (NGOs), Sea Grant, state resource agencies, and universities. Cooperative research projects vary in the level of involvement of each of the participants. Examples include commercial fishermen participating aboard survey vessels; scientists chartering fishing vessels; university and government scientists and fishermen designing new species-specific surveys, and fishermen conducting experiments to improve gear efficiency or reduce bycatch. There is a continuum from cooperative to collaborative research. Cooperative research becomes collaborative research when fishermen are incorporated into all phases of the research process, including formulation of the research question and generation of the hypothesis. In order for this to happen, each participant must understand and appreciate the skills and abilities that other participants possess. Using the scientific method, scientists bring precision, modeling capabilities, statistical verification, and hypothesis generation. Fishermen bring experience on the water and repeated observations of fish and their habitat that can be used to generate hypotheses.

While there has been increased interest in recent years, cooperative research is not a new idea or practice in fishery science. As fishery science developed as a profession in the United States from 1900 into the 1950s, the Commission on Commercial Fisheries and its successor, the Bureau of Commercial Fisheries (both predecessors to the National Marine Fisheries Service), worked with fishermen and their vessels, from shrimp fishermen in the Gulf of Mexico, to the sardine fleet off California, to halibut schooners in Alaska. In the last few decades, cooperative research activities between the fishing industry and the National Marine Fisheries Service (NMFS) have continued and more recently have increased in number and become more formalized. As a result, NMFS requested that the National Research Council perform a study of cooperative research with emphasis on those issues that are essential for the effective design and implementation of cooperative research programs. A committee of experts was assembled to perform a study to address the following statement of task:

> This study will address issues essential for the effective design and implementation of cooperative and collaborative research programs. Cooperative research programs are currently being administered by NMFS to foster the participation of fishermen in the collection of scientific information used in fisheries management. The committee's report will identify design elements necessary for achieving programmatic goals with scientific validity including 1) identification of data needs; (2) setting of research priorities; (3) identification of potential participants; (4) matching of research needs with fishing expertise and access to appropriate vessel and gear; and (5) methods for communicating the findings from this research to the communities involved in or affected by fisheries management.
>
> In addition, the report will address issues essential for the effective implementation of the program such as (1) the maintenance of scientific validity; (2) criteria for awarding and distributing research funds; (3) procedures for evaluating, ranking, and funding research proposals; and (4) ownership and confidentiality of data collected with funds from the cooperative research program. As part of this study, the committee will examine the implementation of existing cooperative research programs.

In conducting this study, the committee reviewed examples of past cooperative research efforts and heard testimony from participants in cooperative research at information-gathering meetings conducted in Boston, Seattle, and St. Petersburg. The committee also heard testimony regarding a number of cooperative research activities in other countries, particularly Canada and New Zealand. The major issues of the report are summarized below.

EVALUATING AND PRIORITIZING
COOPERATIVE RESEARCH

All fisheries research should be evaluated for its potential as cooperative research. If a determination is made that a research project is to be done cooperatively, the relevant question then is what types and degree of cooperative engagement will maximize fishery science and management benefits and are most cost effective. In some cases, only a small degree of engagement may be necessary to generate substantial research benefits from individual research projects. In other cases, the benefits may be smaller in relation to the level of engagement. For relatively minor fisheries research projects (i.e., an expected low payoff), the transaction costs to substantially engage fishermen and other constituents may exceed possible benefits. In almost all of the testimony presented to our committee, though, it was clear that there were potential benefits to cooperative research, and the committee heard testimony about many examples where such benefits had been realized. Therefore, NMFS should explicitly recognize that fishermen and other stakeholder groups can be engaged in many types of fisheries research. The degree of cooperation will depend on regional and fisheries-specific needs and opportunities, the cost of engagement, and the potential gains in achieving science and management objectives.

Three elements critical for evaluating and prioritizing cooperative research include the (1) expected gain in scientific and management benefits; (2) the expected research costs, including opportunity costs for employing fiscal and human resources; and (3) the expected time stream of net benefits (e.g., short- versus long-term net payoffs). Scientists, managers, industry, and other constituents need to collaboratively evaluate potential benefits and costs over time in order to develop consensus priorities for cooperative research.

There is currently no national standard system for setting research priorities, and those who manage the fisheries may have limited input in setting research priorities. This report reviews alternative mechanisms for setting research priorities, including (1) the status quo, (2) NMFS, (3) coordination by regional fishery management councils (FMCs), (4) industry, (5) neutral third parties, and (6) regional research boards. The strengths and weaknesses of the various models for setting priorities were considered.

NMFS fisheries scientists and other constituents should engage fishery managers in strategic discussions for establishing quantifiable management objectives that can be used for prioritizing and evaluating cooperative fisheries research. NMFS must establish an effective and transparent "coordinating" process for prioritizing cooperative fisheries research. Although such a process may be led by NMFS science centers, FMCs, industry, or neutral third parties, NMFS should give serious consideration to establishing independent regional research boards for not only prioritizing and coordinating cooperative fisheries research but also developing innovative and incentive-based research programs, communicating ideas and results, and evaluating research projects.

EXPECTATIONS FOR COOPERATIVE RESEARCH

Cooperative research must meet high scientific standards and also standards of practicality, cost effectiveness, and utility. Peer review at the time of proposal and project completion is essential. Cooperative research should meet the same standards as traditional directed research but at the same time should not be held to a higher standard.

The expectations (diversity of standards) of cooperative research are broader than those of most research projects. Successful execution of such projects must not only include results that are useful to the management process but must also include cross-training of participants, openness, and mutual respect and trust between the scientists and fishermen. These extra requirements are among the factors that make cooperative research more time consuming than other forms of research.

ALLOCATION OF FUNDING

One of the key elements in the statement of task is to make recommendations about how cooperative research funds should be allocated. The default option is that whoever has the money in hand, be it an NMFS regional office, an NMFS regional fisheries science center, or another organization such as a state Sea Grant, has the power to decide. As an alternative, Congress and NMFS should give serious consideration to establishing and funding regional research boards to prioritize and coordinate the use of dedicated funding (earmarks and line items) for cooperative research projects in the region. In addition to these duties, the regional research boards would evaluate NMFS-dedicated research projects for their potential as cooperative research, to foster communication of research results, and evaluate cooperative research projects and programs.

Cooperative research funds must be allocated through a competitive review process where the benefits and costs of performing the research are considered. Some funds need to be made available for rapid response, seed money, and administrative overhead. These functions could be administered by the regional research boards.

LEGAL AND ADMINISTRATIVE ISSUES

There are a number of legal issues that need to be addressed for successful functioning of cooperative research, including Coast Guard licensing and inspection, fishery permits, charter agreements, and insurance. It is recommended that fishing vessels used for cooperative fisheries research by NMFS should meet all U.S. Coast Guard requirements for operation and manning so as to ensure safe operations; that NMFS ensure that appropriate liability insurance is secured for all cooperative fisheries research activities and participants involved in cooperative research prior to the onset of the research project; that NMFS streamline and standardize all permitting procedures for conducting cooperative research projects; and that NMFS and operators of commercial fishing vessels use comprehensive contracting procedures so as to minimize confusion and maximize opportunity for all fishermen to participate in cooperative research.

Cooperative research works when scientists and fishermen realize that the other parties bring valuable tools and experience to the objectives of the project. A prerequisite of successful cooperative research is that all participants thoroughly understand that they are involved in scientific research. Cooperative research must meet scientific standards if it is to be useful to management decision making. Expectations, requirements, and procedures, including the development of agreements carefully detailing the responsibilities of all participants, should be clarified at the beginning of every project.

MAKING COOPERATIVE RESEARCH WORK

Cooperative research will be most successful under one of the following situations: (1) there are perceived threats to the operation of the fishery; (2) there are potential increases in yield to fishermen to be obtained from new information; (3) the fishing industry believes that some form of "accepted scientific wisdom" is wrong and can be shown to be wrong by a well-designed study; or (4) the research fills a significant data gap acknowledged by the participating scientists and stakeholders.

Common problems with cooperative research include (1) poor project oversight or coordination, (2) research fishing achieving its own economic importance beyond its scientific value, (3) research results being leaked prior to peer review and evaluation, and (4) not enough resources being allocated toward the administration and analysis of the project. Cooperative research projects are more likely to be successful when particular attention is paid to project team composition, peer review of proposals, project results and research granting programs, and establishing and employing data verification and quality assurance mechanisms. For larger projects the committee recommends the formation of advisory committees with broad-based membership to facilitate the research and increase the utility of project results. Any use and dissemination of interim project results should be agreed to by all participants at the outset of any cooperative research project; however, at the time of project completion, results should be in the public domain.

For scientists involved in cooperative research, there is an additional time investment and they may not have the same opportunities for publication as those involved in directed research only. NMFS scientists involved in cooperative research need to be given equal opportunity for professional advancement along with their counterparts who do not participate in cooperative research.

1

Introduction

CHARGE TO THE COMMITTEE AND REPORT OVERVIEW

Overview and Definition

Good fisheries management depends on acquiring high-quality information on an ongoing basis. These data provide the backbone of the science used in regulation. In the United States and around the world, the dominant method for obtaining such essential information has been through centralized, government-staffed research programs. While the data obtained from these programs have been of enormous value, the method is not always the most cost effective and often does not make use of the extensive experience of practicing fishermen. In addition to bringing industry and scientists together to engage in research of significant importance to fisheries management, there may be other side benefits resulting from cooperative research, including improved mutual understanding and trust between the cooperating partners. Recognition of the potential direct and indirect benefits of cooperative research has been rising. For example, in a report to Congress on the National Marine Fisheries Service (NMFS) issued in July 2001, the National Academy of Public Administration recommended that:

> The NMFS Assistant Administrator, in collaboration with regional administrators, substantially expand cooperative programs in the area of research, statistics, and dockside extension services to improve external relations.

Whereas in the past, cooperative research was an ad hoc product of agency individuals and fishing groups working together, NMFS now has specific policies supporting and encouraging cooperative research (NMFS, 2001). There has been an increase in cooperative research activities in recent years, boosted by separate and substantial funding provided by Congress. The recent and likely continued increase in cooperative research activity motivated the current study. The committee was charged with identifying and recommending key elements to meet programmatic objectives, to ensure scientific rigor, and to effectively design and implement cooperative research programs. The complete committee statement of task is provided in Appendix A.

It should be noted that in many ways this is not a typical National Research Council (NRC) report because few studies of cooperative research have appeared in the published literature and what has appeared have been the results of cooperative research projects, not an evaluation of cooperative research. The NRC was asked to address the statement of task using available information. Because of the limited availability of published work, the committee had to rely on information provided through case studies, committee members' own experience, and testimony provided to the committee to address some aspects of the statement of task.

COOPERATIVE AND COLLABORATIVE RESEARCH

The nature and level of cooperation can vary greatly among projects. At one end of the spectrum are projects with relatively low levels of cooperation, such as NMFS chartering commercial vessels for surveys (in which the primary form of cooperation is commercial crews helping in the actual daily operation of the surveys) or fishermen keeping logs of fishing activities. On the other end of the scale are cooperative research projects where fishermen and agency personnel work together in all phases of the project, including development of the research question design of the project, performance of research, analysis and interpretation of results, and communication and dissemination of study findings. These types of projects are often referred to as "collaborative research."

SCIENTIFIC POTENTIAL OF COOPERATIVE RESEARCH

A key element in managing the nation's fisheries is our understanding of the abundance and biology of fish in the ocean and how to reduce the impact of fishing activities on fish and protected species. Some of this information can be collected at the docks (as catch data), while the collection of other information requires going to sea and putting instruments, nets, or other sampling gear into the water. Fishermen are at sea every day as part of their normal work activity, and much of the impetus for cooperative research is to integrate the knowledge of fishermen into the scientific process used to regulate fisheries. For every day government research vessels spend on the water, commercial vessels operate for hundreds of days. Some of the traditional cooperative activities, such as logbooks, are designed to accumulate the information available from commercial activity. Commercial fishermen have other valuable skills and information gained through experience that can also be brought to bear in the scientific process.

Fishermen are experts at the use and modification of fishing gear to achieve specific objectives, such as increasing catch efficiency or avoiding nontarget species. As shown by the management directives stemming from studies such as the West Coast Groundfish Mesh Size Study and the Alaska Seabird Deterrent Study, the efficiency of the national fisheries management system can be enhanced if this expertise is used. Thus, it is not surprising that many research efforts investigating fishing gear design have been done cooperatively. As concerns about bycatch of nontarget fish, marine mammals, and birds have increased, the demand for gear research has grown.

Although fishermen often express frustration that fish surveys include areas where they know few fish are to be found, research surveys are not designed to catch fish efficiently but rather to catch them in a well-specified, repeatable fashion that will provide an index of the trends in abundance over time. Scientific surveys are often designed so that there is more sampling effort in areas of higher fish density, but some effort needs to occur in low-density areas. Scientists bring to the process rigor in experimental design and the ability to synthesize large-scale data that are not available to individual fishermen. While most research surveys around the world are designed and conducted by national fisheries agencies with little or no cooperation from fishermen, recent experience on both coasts of the United States has shown that the expertise of fishermen can be important in making sure that the survey fishing gear is operated as efficiently as designed or that the geographic range of the survey is consistent with the geographic range of the fish.

Data from research surveys form the core of most U.S. fisheries stock assessments, and general broad-based surveys have been in place for 20 years or more over much of U.S. waters. However, as regulatory demands for information require greater precision, there is a growing need for additional targeted surveys on specific species. The current and projected NMFS fleet will never be able to respond fully to these growing demands for research surveys. These additional surveys are likely to be conducted from commercial and recreational fishing vessels, if they are to occur at all.

We see great scientific potential in cooperation between scientists and fishermen. Fishermen bring field experience, practical knowledge, and platforms for collection of data. Scientists bring experimental design, the scientific method, and data synthesis. By bringing together the knowledge and skills of these two groups, the quality, quantity, and relevance of research may be greatly improved.

SOCIAL CONTEXT OF COOPERATIVE RESEARCH

A second benefit of cooperative research lies in building better understanding between science and industry. It is clear that when the fishing industry has little confidence in the science, the political process is used to oppose regulations. For fishermen to have confidence in the regulatory process, they must have confidence in the data and the analysis that is used in developing regulations and management plans. Over the course of its open meetings, the committee heard numerous examples in which fishermen's participation in data collection led to greater confidence in the data, the analysis, and the management recommendations that emerged from the process.

THE OBJECTIVES OF PARTICIPANTS

In the testimony presented to the committee, it became clear that there is a wide range of objectives in projects labeled as "cooperative." For the fishing industry, the objectives are primarily associated with improving management. They see cooperative projects as providing an opportunity to collect more and better data at lower cost, to provide scientifically acceptable information about stock dynamics and status for the assessment process, to become more aware of how ongoing data collection programs actually work, and to be true partners in the process.

Bringing fishermen into the data collection process and increasing their understanding of the scientific process provides benefits to NMFS. Cooperative research projects provide a way for NMFS to obtain additional data, to get access to vessels and time at sea, to take advantage of the practical knowledge of fishermen, and to develop effective partnerships with industry. Universities have also been major players in cooperative research programs, participating directly in cooperative research, sometimes serving as intermediaries between industry and government, and in other cases providing scientific input and credibility for projects led by the fishing industry. Cooperative research projects provide university researchers with funding, training opportunities for students, and an entrée into regional management issues. Retired university, NMFS, and other scientists should also be considered for participation in cooperative research projects. Cooperative research provides opportunities for these scientists to continue to make contributions to fisheries science and management.

Recreational fishermen can benefit in ways similar to commercial fishermen and should be considered for cooperative research projects that are appropriate to the vessels, gear, and expertise they possess. The long-standing billfish tagging by recreational fishermen is an example of potential cooperative research.

The emergence of nongovernmental organizations with expertise in marine research, facilitation, or fishing community outreach provides a further opportunity for successful partnerships. Not only can these organizations participate directly in cooperative research projects, they can also serve as an intermediary between scientists and fishermen. An aquarium or independent research facility or organization, often with a reputation for neutrality within the community and an entrepreneurial culture that allows it to act as an interface, can provide flexible services in a way not easily accomplished by government or academic institutions.

Indigenous groups throughout the United States have a major interest in fisheries regulated by NMFS, particularly the salmon fisheries of the West Coast. Indigenous organizations are intensively involved in co-management of many salmon resources and conduct significant scientific research, often in conjunction with state agencies but also with NMFS. The committee heard of no cases where indigenous groups were funded through specific cooperative research grants, as most of their money comes from other sources, but there is nothing to exclude indigenous groups from becoming involved in the cooperative research programs. However, the committee felt that the special nature of indigenous treaty rights and salmon was not a model for other cooperative research activities, and we have excluded this work from our considerations.

Finally, environmental organizations represent another potential cooperative research partner. Although they have not yet been as intensively involved in cooperative research projects as other parties, there are some notable exceptions. Cooperative research projects could provide an increasingly important forum for environmental organizations to participate in the management process and to help solve problems of particular concern.

2

Experience with Cooperative Research in the United States

Cooperative research is not a new idea or practice in fisheries management. State and federal cooperation, collaboration between universities and government, and information collection by fishermen all have long histories. Cooperative research with fishermen as a priority activity and as a tool to increase constituent support for agency programs has received increased emphasis since the mid- to late 1990s.

INTRODUCTION AND BRIEF HISTORY

Cooperative research has a history that parallels that of fisheries research and has evolved along with the agencies and institutions of fisheries science. One of the first fish research vessels was the *Grampus,* a fishing schooner converted in 1886 to the work of Spencer F. Baird and his fledgling research station at Woods Hole, Massachusetts. Although Baird and the Commission of Fish and Fisheries acquired a government vessel dedicated exclusively to fisheries research in 1882, the management agency in its early days relied on information provided by fishermen as well as that gathered by its own scientists. According to National Oceanic and Atmospheric Administration (NOAA) historian William Royce, Baird gathered nongovernmental scientists and naturalists to join him in studying fish and fishing at the U.S. Marine and Biological Laboratory in the 1870s. While scientists pursued research in natural history and biology, the commission described the fisheries themselves. As fisheries science developed as a profession from

1900 into the 1950s, the commission and its successor, the Bureau of Commercial Fisheries, continued to work with fishermen and their vessels from shrimp fishermen in the Gulf of Mexico to the sardine fleet off California to halibut schooners in Alaska.

One of the early programs was a joint research effort between scientists at the Southeast Fisheries Science Center in Miami, Florida, and recreational and commercial fishermen who partnered in tagging sailfish, blue marlin, white marlin, swordfish, bluefin, and yellowfin tuna. From its inception in 1954 to the late 1990s, the program engaged more than 34,000 participants. In Alaska there is a long history of partnerships with industry, beginning with vessel charters as early as the 1950s, and expanding later to surveys and gear research. As fisheries stopped developing and expanding in the 1980s, and as managers imposed more stringent regulations, cooperative projects on gear development, biological surveys, observer programs, and catch statistics continued with renewed importance.

One example of an early effort at gear development was the turtle excluder device (TED). Requirements to include TEDs in southeast shrimp trawls for protection of sea turtles were imminent in the late 1980s. A TED developed by the National Marine Fisheries Service (NMFS) appeared to be effective in excluding sea turtles from trawls while retaining the majority of harvested shrimp. Sea Grant conducted extensive demonstrations of this gear with fishermen, only to be met with opposition. The gear proved to be too cumbersome. To solve this problem, Sea Grant conferred with industry leaders, seeking a more acceptable device from within the fishery. Various devices had been designed for the exclusion of jellyfish that sometimes clogged nets and made it impossible to pull trawls. Sea Grant worked with fishermen to modify these jellyfish excluders for application to deflection of sea turtles from trawls. Funding was appropriated through various Sea Grant programs to evaluate their efficiency. Using the University of Georgia's *RV Georgia Bulldog*, pair towed evaluations were conducted in an area with heavy concentrations of sea turtles. Environmental organizations, commercial fishermen, and government personnel participated in the investigations. A number of trial TEDs were shown to be highly effective in excluding turtles from trawls. Sea Grant then conducted extensive outreach to include numerous demonstrations of the prototype TEDs aboard commercial vessels during shrimp operations. This gear was ultimately accepted by industry, the environmental community, and NMFS and is still in use today.

The lessons learned from the experience of that controversy were applied in subsequent development of gear to exclude finfish from shrimp trawls. A more collaborative approach that engaged industry from the outset was somewhat successful, though still it took many years. In that case, collaboration among NMFS, the Southeast Fishery Development Foundation, fishermen, conservation representatives, Sea Grant, and state agencies resulted in several designs for bycatch reduction devices that have been tested and brought into development in shrimp fisheries in the Gulf and South Atlantic.

Another early gear development program that fishermen initiated was the invention of the Medina panel to release dolphins from purse seine nets. Ideas for employing gear technology and fishing operations to reduce dolphin mortality were developed by the yellowfin tuna fleet in the eastern Pacific Ocean and eventually were integrated into management measures that govern the entire fishery. Large yellowfin tuna often associate with certain species of dolphins. Tuna purse seine fishermen take advantage of that association by locating dolphins visually and then inspecting the herds (primarily by helicopter) to see if a sufficiently abundant tuna school is swimming beneath them. The tuna and dolphins are herded and captured together in the net, but prior to retrieving the net and the tuna, the fishermen attempt to release dolphins by the backdown procedure, in which the vessel puts its engines in reverse, causing submersion of the corkline at the end of the net due to water drag through the fine-meshed net there (the Medina panel). Crew members stand by in small boats near the net in case their assistance is needed. Most of the dolphins are released unharmed, although some do die during the fishing operation. The backdown procedure is an invention of tuna fishermen, born largely of their own interests in avoiding dolphin bycatch and public concern.

The Inter-American Tropical Tuna Commission (IATTC) is responsible for monitoring the incidental dolphin mortality, studying its causes, and providing training and support to fishermen to encourage adoption of fishing techniques that minimize incidental mortality. Dolphin mortality has been reduced by 97 percent since 1986 (from 133,000 in 1986 to less than 3,300 in 1995). A combination of major and minor technological developments, training in the use of the methodology, conscientious decision making by the fishermen, and regulatory pressure to improve performance have all contributed to the reduction in mortality. This process took many years, has been costly, and was not without some unintended impacts. The resulting restrictions on fishing operations resulted in some economic impacts but have not been severe enough to prevent continuation of a substantial purse seine fishery on tuna associated with dolphins.

However, the refusal of certain tuna companies to purchase tuna with the intentional setting of nets on tuna associated with dolphin herds has had the result that some purse seine vessels increase their setting of purse seine nets on tuna associated with floating objects (as an alternative). This has resulted in an increase in bycatch of a range of animals (Hall, 1998).

The IATTC has contributed to the reduction of dolphin mortality by providing a dolphin safety gear program to the fleet, by conducting workshops, and by providing general education to the fleet of the causes of incidental mortality. The fishermen have contributed to the success of the program through their ideas and experiments on safe fishing techniques and by their careful diligence during fishing operations. This is an excellent example of how cooperative effort has resulted in achieving a goal (reduction of dolphin mortality) but led to other unintended consequences due to the changes in gear or fishing practice.

There were few examples of recreational fishermen participating in cooperative research in the case studies presented to the committee. Recreational fishermen have been key participants in tagging studies. They have conducted the majority of tagging and retrieval studies for billfish. In the Gulf of Mexico and South Atlantic, the recreational and charter industries participated in research on snapper discards and helped establish artificial reefs in the Gulf of Mexico and south Atlantic.

In the remainder of this chapter a series of case studies are presented illustrating examples and aspects of cooperative research in the United States. These case studies represent only a small sampling of cooperative research projects conducted in the United States. The committee did not attempt to provide balance by region, type of research, or type of participants involved. Examples and case studies selected were those that either were familiar to the committee members or were described by presenters to the committee and illustrated important points related to various phases of cooperative research projects along with important lessons learned.

NEW ENGLAND AND MID-ATLANTIC SCALLOPS SURVEYS

Atlantic sea scallops, once one of the highest-value species landed in northeast fisheries, have been overfished since the early 1990s. Intense regulation of the fishery since 1994, combined with large closed areas to protect groundfish, had reduced both the amount of time and the areas that were available to scalloping. As open areas became depleted, further limiting fishing opportunities, fishermen became curious about scallop populations in the closed areas. Attempts by industry to get permits for experimental fishing and by scientists to conduct resource surveys coincided with the NMFS's interest in expanding its own survey capability in the region. The desire for more and better information brought the parties to the table to design a cooperative survey of closed areas.

In 1994 the New England Fishery Management Council limited entry to the fishery, restricted effort through days at sea per vessel, reduced crew size, and increased the gear mesh size. At the same time, four areas on Georges Bank were closed to protect depleted groundfish stocks, eliminating all fishing, including dredging for scallops.

Despite the restrictions on effort, landings continued to decline. By 1996 a stock assessment on Georges Bank and in the Atlantic Bight indicated sea scallops were overfished in both areas and were at a low population level. Further restrictions cut allowable days at sea. As scallopers were moved out of areas on Georges Bank, pressure built on the remaining open areas and on mid-Atlantic populations. As a result, additional areas in the mid-Atlantic were closed to scallop fishing. Scientists and managers predicted that in order to meet the requirements of the Sustainable Fisheries Act, days at sea would have to be cut even further—well below what skippers said would cover the costs of a trip.

Meanwhile, in the groundfish closure areas on Georges Bank, the scallop populations prospered. In the first 20 months of the groundfish closures, sea scallop biomass within those areas tripled. These closed areas became the object of curiosity and scrutiny as scallopers had an increasingly difficult time finding productive beds. As other areas in the mid-Atlantic became depleted and the number of days at sea were cut shorter, fishermen wanted information on the condition and abundance of scallops in the closed areas.

There were several attempts by fishermen to acquire experimental fishing permits that would allow entry into the closed areas, but the applications were judged to have incomplete or inadequate detail and/or scientific methodology. None were approved. A subsequent Government Accounting Office investigation of the permit process revealed confusion and lack of clarity in the expectations of the participants. Eventually, a formal proposal by the Center for Marine Science and Technology (CMAST) of the University of Massachusetts at Dartmouth to conduct dredge surveys of scallops in Closed Area II was approved for the 1998 season.

The industry's objective for the survey was straightforward. It wanted to know if there were sufficient scallops of a large enough size to warrant opening up the groundfish closed areas to scalloping. The research objectives for the Northeast Fisheries Science Center (NEFSC) were more complex. While the NEFSC also wanted an estimate of abundance, using commercial vessels to tow the dredges required a means of accounting for the highly varied "footprint" of the bottom gear. Because every vessel and every tow differ, a way had to be figured out to correlate the data from one vessel with the others to provide consistent, reliable results. Furthermore, in order to extrapolate absolute abundance from estimates of relative abundance, an idea of the efficiency of the dredge tows was needed. NMFS also had to get some idea of how much bycatch of flounder and other groundfish occurred during scalloping.

The CMAST application for an experimental fishing permit was approved to allow "fishing" in the closed area by the research dredges. CMAST provided researchers, the industry provided vessels and dredges, and the NEFSC provided electronic equipment and the survey design. From the science center's point of view, the purpose of the project was to calibrate the guidance and action of the NMFS dredge and commercial dredges as a first step in testing the feasibility of using commercial vessels to do surveys.

Participating fishermen were compensated by being allowed to retain and sell 10,000 pounds of scallops per vessel-trip. Fishermen also did not have to count the days working on the project against their days-at-sea allocation. The NEFSC supported the salary of a postdoc at CMAST, and a portion of the proceeds of the sale of the scallops also went to the University of Massachusetts, Dartmouth. A portion of pooled proceeds was used to cover expenses of the participating vessels and the cost of observers.

The NEFSC worked with CMAST and the scallopers on the design of the survey and approaches for addressing potential problems, including developing criteria for selecting participating vessels. The effort to shape a design that would calibrate the industry vessels with each other, and in turn with the NOAA research vessel, required application of substantial technology as well as the operational knowledge of the fishermen.

The survey went forward in August and September 1998 using six vessels, each sampling 100 stations in a two-week period (one every three square nautical miles). In a series of layered experiments, each vessel performed about 300 tows of 10 minutes each. The first major experiment was to estimate the relative density of scallops in the closed areas and, by calculating the average "footprint" of a tow, to extrapolate to the total area. The second major effort was to study the relative efficiency of the dredge by going back and forth over the same plot multiple times, keeping track of the take of scallops each time and of each pass in relation to the total. The project also looked at the scale and patchiness of scallop beds, the rate at which dragging filled the dredge, and tows designed to determine when the scallop dredge actually stopped fishing during haulback.

Controversy arose during data analysis because there was no agreement at the outset on the timing of peer reviews and release of the data. Another difference of opinion among participants related to estimating scallop density relative to tow efficiency and how a model would be used to adjust for repeated tows over the same area. The difference was an estimate of 40 percent efficiency from NEFSC and 16 percent from CMAST. Translated into the abundance estimate, the gap was between 30 million and 60 million scallops. The scientific and statistical committee of the fishery management council and the stock assessment review committee for scallops reviewed the information and accepted the 40 percent efficiency estimate.

Once the pilot was completed successfully and confirmed that the commercial dredge and the NOAA research survey dredge could be calibrated to show consistent results, the NEFSC moved into the second phase of the project. The objectives were to assess the number and size of scallops in the remaining two areas that had been closed since 1994 (Closed Area I and Nantucket Lightship). The other major objective of the second cooperative survey was to evaluate the amount of bycatch in the scallop dredges.

A joint survey was conducted between August 6 and September 1, 1999, in Closed Area I, a triangle of water approximately 40 miles southeast of Cape Cod, and in the Nantucket Lightship area, a rectangle about 30 miles south of Nantucket. Two scallopers were chosen by lottery to participate in the biomass estimate portion of the survey, and two other fishing vessels were chosen by lottery to participate in the bycatch portion of the experiment. The industry also provided crews. Scallopers used their allocated days at sea and retained for sale 14,000 pounds of sea scallop meats from their 10-day trips, a value of about $80,000. Vessels only had to count days at sea actually used in the survey tows.

As with the 1998 project, the NEFSC was responsible for the survey design, the industry provided platforms and crews, and the Virginia Institute of Marine Studies provided researchers and other scientific assistance.

Conclusions

Results of this cooperative research not only showed an abundance of scallops of large size, but the intensity and scale of information it was able to generate about scallops, their habitat, and the other species associated with them in the closed areas allowed managers to devise bycatch reduction measures that enabled additional openings, resulting in landings of scallop meats worth approximately $36 million. This case study demonstrates how incentives can be built into the collection of information to create positive feedback for the industry and at the same time maintain scientific credibility of the survey methodology and data analysis.

THE WEST COAST MESH SIZE STUDY, ENSURING RIGOR AND EXPERIMENTAL DESIGN

The West Coast Groundfish Mesh Size Study was conducted during the late 1980s through the early 1990s off the coasts of Oregon, Washington, and California. A major impetus for the study was the finding that the predominant management tool being used to regulate the fishery (individual-species trip quotas, or "trip limits") was causing significant and increasing discarding of otherwise marketable fish and that substantial discarding of below-market-sized fish was also occurring (Pikitch, 1986; Pikitch et al., 1988; Pikitch, 1991). The fishing industry advocated that mesh size research be conducted to evaluate whether new mesh size regulations could replace or diminish reliance on trip limits and reduce discard levels. This case study focuses on the development of an experimental design for the field component of the research project conducted in 1988.

At the project's inception, an advisory group was established and included members from industry, scientists and managers from state and federal agencies, and scientists from academia. While a small research team spearheaded the project and managed and performed daily tasks, all major decisions were made via consensus of the advisory group, which met numerous times during the course of investigations. The diversity of interests represented in the advisory group shaped a research agenda designed to meet the highest scientific standards while minimizing costs and maximizing the chance that the results of the study would actually be applied in the management of the fishery. Thus, from the outset, it was clear that the standards applicable to the West Coast Groundfish Mesh Size Study were broader than those applied to typical scientific research projects.

A great deal of time and effort was employed to establish an experimental design that would meet all the criteria needed to meet overall project objectives. An early design constraint agreed to by all participants was that all work would be conducted aboard commercial fishing vessels operating under commercial fishing conditions (i.e., using normal commercial gear deployed under customary conditions and fishing grounds and for usual durations). This design element was selected to ensure realism and, ultimately, the utility of the results for management purposes. Another advantage of the use of these vessels was reduction in cost. Not only did the survey provide data based on realistic conditions by using these vessels, it did so at a much lower cost than if a government research vessel had been chartered.

All parties agreed on the need for scientific rigor. The methods used to evaluate experimental design options were subjected to several levels of peer review, including informal reviews by the investigator's peers in academia and NMFS, formal review by the Pacific Fishery Management Council's (PFMC) scientific advisory committees, and ultimately submission and publication in a refereed science journal (Bergh et al., 1990). Initially, with convenience and fishing efficiency as primary considerations, a design in which each vessel would fish with a single mesh size during a given trip was advocated by the fishing industry. Calculations performed by the scientific team members demonstrated that the number of trips needed to detect a significant difference of the magnitude expected was prohibitively high under the requested scenario. Thus, while this scenario passed scientific muster, it failed to meet other standards of feasibility and cost efficiency. An alternative, albeit less convenient design, was considered whereby each vessel would change mesh sizes every tow according to a predetermined randomized pattern. The number of trips needed under the alternative scenario was an order of magnitude less than that required by the original design. With the advantages of the alternative design far outweighing its disadvantages, it was adopted via a consensus of the parties.

Another key design element was the use of volunteer fishing vessels that were exempt from trip limits while engaged in cooperative research fishing trips. The trip-limit waiver was facilitated by the multistakeholder advisory committee and ensured that fishing trips would be scientifically productive (i.e., not terminated prematurely as a result of attainment of trip quotas) and also served as a powerful financial incentive for fishermen to cooperate. As noted above, the use of volunteer vessels, while demanding much more time and effort to coordinate on the part of the research team than using vessel charters or chartering a government research vessel, greatly reduced the costs of conducting the study. In addition, by involving a much greater fraction of the fleet than would a single-vessel charter or government charter, this design element contributed to both the realism and acceptance of the results by a broad segment of the industry.

The detailed findings of the field studies are provided by Pikitch (1991, 1992) and Pikitch et al. (1990). The results demonstrated that an increase in mesh size would greatly reduce the catch (and hence the discard) of small unmarketable fish. For example, discards were approximately half as numerous for catches obtained using 114-mm (4.5-inch) mesh codends than for catches obtained using 76-mm (3.0-inch) mesh codends (Pikitch et al., 1990). Based largely on this work, the PFMC voted to increase minimum regulated mesh size for bottom trawls from 76 mm to 114 mm in 1991.

Conclusions

First, it is clear that a partnership among stakeholders, including scientists, industry, and regulators (in this case in the form of an advisory committee) was essential to the success of the study. The partnership ensured the scientific integrity, practicality, and cost effectiveness of the experimental design and facilitated ready application of the results to alter management practices. The advisory committee also paved the way for obtaining the needed waiver of regulations and obtaining volunteer vessel participation. The focus on experimental design also proved to be essential. Results of the analyses performed led to adoption of an experimental protocol that made for efficient use of resources and allowed sufficient sampling effort to be deployed. The ready adoption of the results by managers to change regulations was not a "lucky coincidence" but rather a critical objective of the study from the outset which led to the adoption of several key design constraints.

This case study also illustrates the "higher standard" to which cooperative research projects can be held relative to other scientific projects. In the case of the West Coast Groundfish Mesh Size Study, a scientifically valid experimental design was a necessary criterion for success, but that in itself was insufficient. From the outset, practicality, acceptability, cost effectiveness, and utility were key design criteria. In this example, faithful adherence to these criteria in development of the experimental design led not only to a successful research project but also to a rapid and successful management outcome.

WEST COAST VOLUNTEER LOGBOOK PROGRAM

One of the most important elements of any partnership is a clear understanding of expectations among parties. The following case study illustrates that, while cooperation can produce benefits, it can also create significant problems, particularly if it is not adequately funded and suffers from conflicting expectations about the responsibilities and priorities among the parties.

Although the West Coast groundfish trawl fishery adopted minimum mesh size regulations, the fishery continued to discard significant portions of the catch due to economic and regulatory constraints. In particular, high levels of regulatory discarding were occurring in the shelf slope trawl groundfishery due to trip limits and harvest quotas set by the PFMC. Decreasing stock estimates of the slope trawl DTS (dover sole, thornyheads, sablefish) fishery, combined with fleet overcapacity, were resulting in tightening trip limits. Managers assumed that rates of regulatory discarding would change and possibly increase as fishermen responded to these limits. Groundfish allocation issues were also contributing to increasing attention to the trawl fishery and its discard practices. Discard rates used by the PFMC in estimating optimum yields and trip limits were based on a study conducted by Pikitch et al. (1988) during the mid-1980s, before significant reductions in quotas and trip limits. Management believed it was critical to update discard estimates. The PFMC and NMFS's Northwest Fisheries Science Center (NWFSC) began discussing management alternatives, including mandatory observer programs or full retention harvest regulations. In 1995 the Oregon Trawl Commission (OTC), which represents Oregon-based groundfish and shrimp trawlers, agreed to collaborate with the Oregon Department of Fish and Wildlife (ODF&W) to conduct a cooperative pilot program to update estimates of discard rates and test approaches for developing comprehensive at-sea data collection programs.

The primary science and industry cooperators were ODF&W; OTC; Oregon; and to a lesser extent other West Coast shelf and slope trawl vessel owners, skippers, and crews who volunteered their time and effort.

The initial project had three primary objectives: (1) develop approaches to efficiently organize and administer a comprehensive trawl groundfish data collection program; (2) scientifically estimate trawl fleet slope bycatch and discard rates for all federally managed species; and (3) improve onboard biological sampling. Associated with these goals were 19 subobjectives and 34 tasks (Saelens, 1995). An important project criterion was to obtain scientifically valid data using methods that did not interfere with the normal course of fishermen and vessel operations and behavior. A key set of subobjectives was to obtain representative data of the entire fleet by placing observers on over 10 percent of the trawl fleet and enhanced logbooks on 20–40 percent of trawl vessels. A key expectation was that different data sources could be collectively used to improve scientific analysis of catch and discards. These data included: (1) enhanced logbooks maintained by operators of vessels; (2) fish ticket weights generated by processors and collected by shore-based ODF&W personnel; (3) catch and discards data and biological samples collected by observers; and (4) biological samples collected by industry.

Project concepts were codeveloped by ODF&W and OTC. ODF&W developed a proposal that was submitted to OTC for approval. The OTC received permission by its constituents to levy an additional tax of 0.5% on exvessel groundfish revenues to pay for the program. ODF&W hired observers, developed the enhanced logbooks, and developed methods and protocols for collecting data and biological samples. ODF&W was also responsible for collecting and maintaining raw data and obtaining experimental fishing permits. The OTC was expected to help market the program and locate cooperating voluntary vessels, which received no direct compensation except caps and jackets for skippers and crew. It was expected that federal scientists would analyze the data, given that ODF&W did not have adequate scientific staff. The first year of the project (Phase I) was expected to be a learning process in order to improve methods and protocols. It was expected that data and findings would be shared with industry, the public, and managers during the course of the study.

Difficulties during Phase I exceeded managers' expectations. The first proposal by ODF&W was rejected by OTC because of perceived high costs. As a result, the project was rewritten as "bare bones," which retained most objectives but cut staff support. Managers received additional funding from ODF&W and the NWFSC, but the funds were used to expand the project to the shelf trawl fisheries and for increased participation from California and Washington. No additional funds were used to support administrative or analytical staff or to compensate participating vessels. Project managers and the OTC encountered problems enlisting the cooperation of an adequate number of vessels, and there was confusion regarding the respective responsibilities of OTC and ODF&W. This problem compounded efforts to maintain regularly scheduled trips and to fully employ scientific observers. Turnover among observers was high, which resulted in increased administrative costs.

By 1996, stock problems were increasing among some shelf groundfish species, which expanded focus on the shelf portion of the study. Project managers were being pressured to release preliminary data. Managers, however, were reluctant to provide data, given the project's problems, limited number of observations, and lack of scientific review and analysis. ODF&W then received complaints that project managers were deliberately withholding findings for political reasons. In response, project managers felt compelled to release summaries of the raw data at public meetings that received significant press coverage. Rather than resolve scientific questions, the release of data that were not scientifically validated heightened management tensions. Although the project limped along for the next year and a half, there was little management support. The project never achieved its vessel coverage objectives. There was also no final report published. Although some of the data have been used by NWFSC scientists to update discard rates for selected species, collectively the project data could not be used to scientifically determine trawl discard rates for all federally managed shelf and slope species. Some of the administrative and data collection methods developed during the study were integrated into the federally funded groundfish observer program instituted in 2001.

Conclusions

The data enhancement project was designed to address a politically contentious data and science problem. The OTC and ODF&W demonstrated leadership in attempting to address this problem. The study did produce useful administrative methods and protocols, and generally positive relationships were experienced between fishermen and observers. This helped smooth the transition into a newly instituted mandatory observer program (Bernstein and Iudicello, 2000). However, the project did not achieve many of its objectives. The project's administrative problems heightened political tensions and created the impression, however false, that the project was designed to produce biased data in favor of the trawl fishery. In retrospect, it is clear that the project was overly ambitious, underfunded, and suffered from conflicting expectations about responsibilities and priorities.

REDUCING SEABIRD BYCATCH IN THE
ALASKA LONGLINE FISHERIES

The participation of environmental groups in cooperative research can serve to provide momentum for policy choices based on results and for promoting similar research in other regions. Following successful conclusion of a seabird bycatch study in Alaska, the Audubon Living Oceans Program helped initiate a similar project in Hawaii.

The incidental mortality of seabirds in longline fisheries is a serious conservation issue worldwide. In Alaska the presence of an endangered species of seabird, the short-tailed Albatross (*Diomedea albatrus*), heightens the importance of that issue. According to the U.S. Fish and Wildlife Service's (USFWS) Biological Opinion, any mortality over six short-tailed Albatrosses within a two-year period (two in the halibut fishery and four in the rest of the groundfish fisheries) would trigger a Section 7 consultation mandated by the Endangered Species Act. The result could be an interruption or even a closure of Alaska's $300 million longline industry. The Biological Opinion requires that mitigation measures be used in these fisheries and that research be done to test the effectiveness of these measures. In 1996 the North Pacific Fishery Management Council (NPFMC) passed regulations mandating certain seabird deterrent techniques for longliners. Then, in 1999, Washington Sea Grant began a cooperative research project that resulted in recommendations to refine and improve these mitigation measures. The revised measures then became required conditions for longlining, as per new regulations subsequently passed by the NPFMC.

This cooperative research project lasted two years (1999 and 2000) and involved two fleets: the catcher-processor fleet targeting Pacific cod and the catcher-vessel fleet fishing under individual fishing quotas (IFQs) for sablefish. Along with Washington Sea Grant, three other institutions collaborated on this project: the University of Washington, USFWS, and NMFS. Two fishing associations also collaborated: the IFQ vessels all were members of the Fishing Vessel Owners Association, and the Pacific cod vessels all were members of the North Pacific Longline Association. Working fishermen collaborated in two ways: they identified possible deterrents at the beginning of the project and decided along with the program director which of those would be most suitable for testing, and they tested these deterrents on their vessels using experimentally rigorous tests while actively fishing under typical conditions. Observers trained and certified by the NMFS collected data aboard the participating vessels.

The use of commercial vessels in the seabird deterrent survey was in the tradition of the International Pacific Halibut Commission (IPHC) and its use of this kind of vessel for halibut stock surveys. Rather than incur the investment of its own vessel and its maintenance and operations, the IPHC has for many decades chartered working fishing vessels as a more cost-effective way of having an oceangoing platform from which to gather data in a standardized format.

Conclusions

The use of working fishing vessels accomplished two goals. Not only did it provide a realistic setting for the testing of bird deterrents, it also kept costs down. No money was spent on vessel charters. The incentive for fishermen to participate was to receive free observer time, something they normally pay for. All the government had to pay for was the cost of each observer, a small fraction of a vessel charter, and a much smaller fraction of a government research vessel charter. In this regard, this research was similar to the West Coast Groundfish Mesh Size Study.

The initial seabird deterrent regulations passed in 1996 by the NPFMC derived from anecdotal information from fishermen and from seabird deterrence regulations from other parts of the world. The results of the cooperative research in 1999 and 2000 clearly identified the most effective techniques for deterrence and also showed that some of the former regulations were ineffective. The resultant regulations are both simpler and more specific than the previous regulations. For a full report of this research, see Melvin et al. (2001).

COOPERATIVE FINFISH RESOURCE ABUNDANCE SURVEY IN THE MID-ATLANTIC BIGHT

Communication is important at all stages of cooperative projects, particularly in creating clear expectations at the outset. A critical aspect is when and how results of cooperative projects are communicated. A successful project in the Mid-Atlantic Bight was put at risk by premature release of information.

In a January 2001 meeting, Rutgers University suggested to NMFS that fishing vessels towing alongside research vessels in the finfish resource abundance surveys might be able to provide supplemental data on diversity, age composition, and abundance of fish within the survey area. An ancillary benefit of the proposed cooperative research was that fishermen would have more confidence in the NMFS survey if they were involved in the research and were able to compare the catch data and observations on their vessels with those on the research vessel when towing "side by side." A survey protocol was developed by Rutgers University and the NEFSC.

The first cooperative survey was conducted and resulted in the completion of 37 side-by-side tows. A report was prepared by Rutgers University and released in August 2001 after the NEFSC completed its peer review. The report stated that side-by-side tows could be an important part of the survey in the future but that the observations of a single experiment must be repeated in multiple experiments before the results could be considered reliable. A second side-by-side survey was planned for the fall 2001, and as a result, 59 additional side-by-side tows were scheduled for September 2001. A report was submitted to NEFSC by Rutgers University in January 2002.

During 2001 a newspaper reporter interviewed fishermen and scientists participating in the cooperative survey prior to the official release of all data and conclusions and wrote a critical article. In 2002 attempts to arrange for additional cooperative surveys (this time as augmentations to the NMFS survey) failed when fishermen were informed that the NMFS spring 2002 trawl survey had already commenced. Finally, in August 2002, representatives of the fishing industry, Rutgers University, and the NEFSC agreed to perform further cooperative survey efforts in 2003. Specifically, it was decided that side-by-side tows would not be as useful as survey augmentation.

Conclusions

The case study illustrates how the premature release of information to the media can erode trust and put cooperative research efforts at risk of failure. The basis of a cooperative research project is trust, and it must be earned by all participants in the project.

3

Examples of Cooperative Research in Other Countries

INTRODUCTION AND PERSPECTIVE

Other countries have also explored cooperative research, often in a different institutional setting from U.S. fisheries. The committee heard testimony regarding various cooperative research projects in eastern and western Canada and in New Zealand. This chapter presents an overview of these experiences, including the essential elements of a range of cooperative research ventures. Conclusions that may be applicable to cooperative research in the United States are also provided.

SENTINEL SURVEYS: AN OVERVIEW OF SENTINEL FISHERIES IN EASTERN CANADA

Cooperative research between Canada's Department of Fisheries and Oceans (DFO) and the fishing industry of the east coast of Canada is in the form of cooperative surveys, mainly as sentinel surveys or sentinel fisheries, which are limited commercial fisheries designed to maintain a continuous record of fishery-dependent data during an otherwise closed period. The primary impetus for development of cooperative research on the east coast of Canada was the closure of most of the major cod (*Gadus morhua*) fisheries in 1992–1993. The Fisheries Resource Conservation Council recommended that DFO put in place programs of "sentinel surveys," or sentinel fisheries, to ensure the continued collection of information for the stocks during the fishery moratorium. DFO responded by implementing a number of such sentinel surveys from Newfoundland and Labrador to the Scotian shelf. Most of these were initiated between 1994 and 1996. These surveys were restricted in their total removals from the stocks and followed scientific protocols for data collection. The vast majority of these initiatives are funded by the Canadian government, both through deployment of government science staff to design and implement them and through payments to individual fishermen and their boats to execute them. The overriding principles of these surveys are to determine the overall abundance of cod in specific areas or stocks, often in areas that government trawl surveys could not access (inshore waters and untrawlable bottom), and to involve the fishing industry directly in the scientific assessment process. The surveys utilized designs ranging from the use of local knowledge and expertise to monitor local catch rates to formal design with area stratification schemes and consistent survey protocols for all participating vessels to estimate stock biomass trajectories. In addition to these primary objectives, most of the surveys have ancillary objectives, including determining population age structure, fish condition, spawning times and locations, and predator-prey interactions.

From 1995 to the present, the Canadian government has invested between 6 million and 7 million dollars annually for a total of about 27 distinct sentinel surveys on the east coast of Canada from Newfoundland and Labrador to Nova Scotia. These surveys involve some 19 fishing industry organizations in all five eastern Canadian provinces and represent approximately 30 percent of the budget that the DFO allocates to the stock assessment process. Given this related high cost, a comprehensive review was undertaken in 2001 with a mandate "to provide recommendations designed to position the sentinel fishery (survey) program as an effective and cost efficient component of the groundfish assessment process on a longer-term basis" (Workshop on the Groundfish Sentinel Program—2001). Although the primary objective of this review was not strictly comparable to the statement of task for this committee, a significant number of the ancillary findings may provide useful guidance.

The following summarizes the conclusions of the review:

- Standardized information collection, processing, and storage protocols, with clearly defined responsibilities for each partner, offer the most successful approach.

- Data quality assurance through use of onboard observers increases the overall acceptability of the information gathered.
- The degree of collaboration in the analysis of survey information was highly variable, with some fishermen partners contributing significantly and others relying on the science partner. Fisherman partners often expressed a desire or willingness to be involved in the analysis of the information.
- In the majority of cases where cooperative sentinel surveys have been put in place, these have augmented existing data available for the assessment of individual resource status rather than leading to new analytical approaches. In some cases, there were data collected during the cooperative surveys that provide new insight into other aspects of the target populations or into the marine ecosystems in which they occur that could lead to new analytical approaches.
- Once implemented and established, cooperative surveys (or other cooperative research arrangements) lead to the development of additional cooperative research projects of interest to all partners.

There is a tension between the rigorous scientific design and adherence to predefined protocols demanded by scientists and the more adaptive "sizing up" approach used by fishermen to determine resource status. This is an important area of discussion and mutual compromise between the partners. Achieving a workable balance between fishermen's expertise and a defensible statistical design is essential for the effective implementation of cooperative surveys. The discussions leading to this compromise are most effectively achieved through a process of coeducation. It is through these discussions that ancillary objectives of the survey are established.

As an example, the participants in the Scotian Shelf and Southern Grand Banks Halibut survey, which was established to provide a number of indices of stock abundance and condition, soon realized that the observers onboard their vessels and the scientific partners provided a good opportunity for them to address a long-standing problem of live to processed weight conversions. The primary issue was that the fisherman considered the conversion ratios used by fisheries managers to estimate live weight (catches) from landed weight (in a number of processed forms) were incorrect, resulting in an overestimate of live weight removed from the population. The study was designed, carried out, and published and resulted in a change of the official conversion factors, with a resultant increase in available revenues to the fishing community (Zwanenburg and Wilson, 2000).

The cost of cooperative surveys (as implemented on the east coast of Canada) is relatively high and increases the overall costs of resource assessment. However a review of cost benefits for the surveys was inconclusive because it was not possible to quantify either the degree or value of improvement to the assessment process. It was also concluded that there was an array of intangible benefits derived from the surveys (increased dialogue between fishermen and scientists, improved understanding of assessments by fishermen) that are difficult to quantify.

Changes in the design, implementation, and analysis of cooperative survey data are continually proposed by both partners and are indicative of a healthy debate and an open dialogue. There are few voices calling for the elimination of these surveys.

Cooperative surveys are likely to become an integral part of the future fisheries on the east coast of Canada. They will become part and parcel of "doing business," and their associated costs are likely to become part of the overhead for the fishing industry.

These surveys do not represent an exhaustive list of all cooperative research projects that have taken place or are currently in place for the east coast of Canada; however, they likely represent a majority. There are also a number of cooperative surveys and research initiatives that do not rely as heavily on government funding.

GRAND BANKS—SCOTIAN SHELF ATLANTIC HALIBUT LONGLINE SURVEY

Unlike most other commercially exploited demersal fish, Atlantic halibut are not a schooling species, making it difficult to derive fishery-independent estimates of population abundance from standard groundfish otter trawl surveys. In 1998 a cooperative survey to develop a more reliable index of abundance of this species was developed and initiated as a cooperative venture between expert halibut fishermen and DFO (Zwanenburg and Wilson, 2000).

Process and Discussions Leading to Establishment of the Survey

Although the need for the halibut survey was evident, the process by which to design and implement it effectively and successfully was less evident. From the outset, DFO agreed that the fish caught during the survey would be used as payment to the participants for conducting the survey. In addition, to avoid conflicts with nonparticipants, halibut caught by the survey was not counted against the total allowable catch (TAC) but was treated as an annual overrun of the TAC. These were key concessions in that halibut has a very high landed value relative to most other commercial fishes ($Can 7.00–13.00 per kg landed) so that the revenue generated by even small (by industry standards) catches would be sufficient to make participation in the survey economically feasible. Allowing the catches to be taken as an overrun of the TAC prevented a backlash from other non-participating fishermen. Since the overrun, at 10 percent of the TAC, is small, its potential impact on the viability of the population is outweighed by the increase in knowledge.

Even though the proper climate existed between fishermen and government scientists, there were differences of opinion and rivalries among the participating fishing communities. The fishermen, however, were united in their opinion that DFO should not have too much control over the survey and that fishermen themselves should be managing all but the scientific aspects of the survey. This ownership of the process eventually united fishermen and allowed them to work together despite regional rivalries. The outcome of these discussions was that the fishermen themselves would manage all the logistics and finances within a separate organization and that the DFO scientists would be responsible for ensuring that the design of the survey would generate scientifically defensible information.

DFO regional offices were established to oversee the survey. They developed application forms to ensure that each applicant met the criteria established for participants during the initial deliberations. These criteria included: (1) vessel size and safety certification; (2) fishing history to document expertise in the halibut fishery; (3) a check of fishing violations history; and (4) a requirement to sign a contract stating a willingness to adhere to survey protocols, including carrying onboard observers for help with data collection and data verification. Vessels were also required to contribute 2 percent of their gross halibut catch value with the coordinator upon settlement of each trip. This fee was applied to all participants, both to pay for the costs of coordination activities and to ensure that no participant would lose money as a result of participating in the survey. It was agreed that these tax monies would be used to redeem at least the money spent (outlay) in participating.

A selection board consisting of an independent chair, a representative from DFO (nonvoting), and representatives from local fishing communities and other community organizations was used to select participating vessels for the halibut survey.

Objectives of the Halibut Survey

The overall objective of the survey was to develop indices of halibut abundance and to increase the overall knowledge base for this species. There were a number of ancillary objectives, including the collection of information on conversion factors between processed and live weights of Atlantic halibut, on bycatch, and on oceanographic conditions. Information on predator-prey relationships would be gathered to allow for a more "ecosystem-based" perspective in the determination of halibut stock status. In addition to improving the knowledge base for this species, it was intended that the interaction between fisheries scientists and this sector of the fishing industry would develop effective working relationships between the two groups and, on the part of the fishermen, engender a sense of ownership of the resource and a sense of responsibility for its sustainable use in an ecosystem context.

General Design

To satisfy both the desires of the fishermen participants to contribute their knowledge and experience in determining halibut abundance and the necessities of statistical rigor, the halibut longline survey was designed as two phases. The first is a fixed station phase using historical catch rates as the stratifying variable, and the second is a commercial index fishery. The fixed station phase was designed to give an unbiased annual estimate of halibut abundance, and the commercial index phase was designed to allow participating fishermen to contribute their knowledge and fishing skills in developing an annual standardized estimate of commercial catch per unit of effort.

In the fixed station phase of the halibut survey, the fishing locations and protocols were strictly prescribed (fishing location, number and size of hooks, set time, soak time, etc.). All of these sets were to be observed by a certified independent observer. Each participating vessel agreed to complete a predetermined number of these sets in return for which they would be allowed to fish a number of "commercial index" days. The only restrictions on the participants during the commercial index phase were a maximum of 7,000 hooks per day and exclusion from existing closed areas.

Detailed information on catches and biological samples was collected for both phases of the survey.

It was agreed that in 1998 all survey activities would include the presence of certified onboard fisheries observers. This requirement was to ensure that an independent observer could verify information collected by the survey and that the requisite biological information would be collected for all survey activities. The requirement for observer coverage was reduced in subsequent years such that at present all fixed station survey activities are observed, but only a smaller percentage of commercial index sets are observed. Captains of participating vessels are responsible for arranging and paying for observers directly.

Each year, following the completion of the survey and data editing, results of the survey are presented in meetings with all participants. The results are in the form of maps showing catch rates for both Atlantic halibut and other species of interest and estimates of fixed station and commercial catch rates. Feedback to these presentations is extensive and includes detailed accounts of anomalous observations and ancillary information not formally included in the data collection protocols.

Conclusions

Although the full value of the halibut survey is just now being realized, it has already been immensely successful in increasing the knowledge base for this species and in fostering an effective working relationship between halibut fishermen and fisheries scientists. Keys to the success of this initiative to this point are:

- The high landed value of the target species and the use of catch revenues to cover the costs of the survey.
- The degree of responsibility assumed by the industry participants, particularly in the management of logistical and financial aspects of the survey.
- The agreement on survey design, protocols, and data flows and storage prior to starting the survey, which included compromises by all partners to ensure an acceptable overall product.
- Regular feedback of results to participants on an ongoing basis.
- A willingness by both partners to commit to a relatively long-term project.

In order for the survey to become a reliable indicator of stock abundance, it is essential that the survey be put in place with a view to long-term sustainability. A survey such as this begins to pay dividends as an indicator of abundance only when it has been in place for a number of years, with a consistent design. Significant changes in design from year to year will erode the efficacy of the survey as an index of abundance. It was therefore important that the initial design agreed upon by the participants be carefully planned. The initial frame agreed upon for the halibut survey was 5–10 years.

Both the scientists and the fishermen participants realized that a long-term commitment was essential to the success of this venture. First, it is not feasible or appropriate to interpret the results of an abundance survey carried out for only a few years. Although it was difficult to convince the fishermen partners of this idea initially, it has now become accepted, and an interim target of five years before significant interpretation of results has been adopted. Commitment beyond this initial period is essential in that there exists no viable alternative fishery-independent method of monitoring the halibut resource. Trawl surveys are ineffective at estimating halibut abundance and it is unlikely that the government will develop an extensive longline survey, even if it were financially viable (estimated cost is between $750,000 and $1 million). In the absence of this survey, the fishermen would once again be without an effective voice at the management table, and the management body would be forced to adopt a highly precautionary approach to establishing (lower) allowable catch levels in the absence of effective monitoring information. This factor, coupled with the high landed value of the species (making the operations of the survey economically viable) improves the prospects of its long-term maintenance.

FISHERMEN AND SCIENTISTS RESEARCH SOCIETY IN EASTERN CANADA

There has long been recognition among fisheries scientists and fishermen that a key to improved management of fisheries is better communication between these two groups. There have been numerous attempts (especially since the establishment of Canada's exclusive economic zone in 1977) to improve and broaden such communications. The closure of the cod fisheries on the eastern Scotian shelf (and the effective closure of the remaining fisheries because of cod bycatch restrictions) provided an incentive for each of these groups to open a meaningful and effective dialogue to share information essential to the long-term sustainability of these fisheries. This dialogue resulted in the development and implementation of the Eastern Scotian Shelf Sentinel Survey for cod and other surveys like the halibut survey outlined above and the establishment of the Fishermen and Scientists Research Society (FSRS).

The FSRS was formally established as a nonprofit organization in January 1994 after a year of discussions between fishermen and a small number of fishery scientists. The objectives of the society are to promote communication among fishermen, scientists, and the general public and to establish and maintain a network of fishermen and scientists capable of conducting cooperative research and collecting information relevant and necessary to the long-term sustainability of marine fisheries. The society was formed out of the recognition by both fishermen and scientists that each had valuable contributions to make to the long-term stewardship of living marine resources. A partnership based on effective communication and common goals was a necessary prerequisite to realizing this objective.

The early days of developing the society involved lots of discussions in kitchens, town halls, church basements, and bait sheds to build initial bridges and trust between fishermen and scientists, develop some common language, and negotiate common goals. These early steps were necessary to overcome the significant mistrust that had developed between the two groups preceding the fisheries closures. Many fishermen felt that scientists had nothing to offer because they were not fishermen, and many scientists felt that fishermen, without formal training, could not meaningfully participate in scientific discussions about fish stocks. From these humble beginnings, with not much more than a willingness to talk and a feeling that cooperation was better than confrontation, the society evolved. It has now developed into an effective organization that brings the knowledge of fishermen into the scientific arena by agreeing on rules of information and that educates fisheries (and other) scientists by making them aware of the wealth of knowledge about fish and fishing that fishermen gain by experience.

The society was officially incorporated as a nonprofit society in 1994, with the following goals:

- Promote further cooperation between fishermen and scientists.
- Collect and interpret accurate information.
- Protect fish stocks and marine ecosystems.
- Contribute to the establishment of a more sustainable fishery.
- Contribute to the viability of coastal communities.

- Become a financially self-sustaining, nonprofit organization over the long term.
- Further educate all involved in the society.
- Perpetuate livelihoods as fishermen.

The formal objectives of the society were debated and finalized at the first annual general meeting (1994) and are:

- To establish and maintain a network of fishing industry personnel to collect information, for use by members of the FSRS, relevant to the long-term sustainability of the marine fishing industry in the Atlantic region.
- To facilitate and promote effective communication among fishermen, scientists, and the general public.
- To participate, as appropriate, in research projects of other agencies and institutions that require the collection of information relevant to fisheries and marine environmental monitoring.
- To generate revenue, where possible, from activities related to information gathering, sample collection, and environmental monitoring to promote the continuation of the FSRS.
- To analyze and disseminate information generated through the activities of the FSRS.
- To facilitate the provision of training to members of the FSRS as may be necessary or desirable in carrying out the objectives of the FSRS.
- To avoid, by action or inaction on the part of the FSRS, the perception that the FSRS is a lobby group representing the interests of either the fishing industry or the scientific community over that of the long-term sustainability of the fishery as a whole.

It must be emphasized that the FSRS is not a fisheries management body or a lobby organization and will not represent the interests of one fishing group over another. The society does, however, inform management bodies through the information it collects and the research it undertakes. As such it has provided significantly improved information for both finfish and invertebrate resource management.

The society has also formulated a code of ethics to which all members have agreed to adhere. The code states that:

- Members must collect and report as much relevant and accurate information as possible, according to instructions and requirements.

- All information and data collected under the auspices of the society remain the property of the society, not withstanding any access granted to individuals for interpretation and analysis.
- Conclusions presented in any products other than society products must include a disclaimer to the effect that the interpretation and conclusions reached by the person preparing the product, be he/she a member, nonmember scientist, or other group or individual, when analyzing the society's data are not necessarily those of the society.
- Members, nonmember scientists, or other groups or individuals given permission to analyze the society's data must first present their conclusions to the society. Members, nonmember scientists, or other groups or individuals analyzing the society's data must acknowledge the society as the source of the data.
- Members shall communicate, educate, and promote, wherever and whenever possible, the nature and importance of the responsible commercial fishing industry in Atlantic Canada.
- Members shall not compete with the society for contracts.
- Members shall constantly act with fairness and integrity in dealing with clients and employees.
- Members shall conduct themselves in a professional and dignified manner and relate to others with courtesy and respect.
- Members must declare any potential conflict of interest, whether real or perceived.

At present the society has over 200 members throughout Atlantic Canada and elsewhere. The society manages a comprehensive annual survey of fishes on the eastern Scotian shelf and is involved in a wide range of research projects in collaboration with the DFO, nongovernmental organizations, and universities in the region. Projects include collection of detailed information on fishing practices, catch rates, species composition of catches, fish condition factors, and information on fish diets essential to understanding their roles in marine ecosystems as a whole. The society is also conducting studies to determine levels of lobster recruitment (the number of young lobsters produced each year), seasonal changes in lobster weights, and whether lobsters from different areas grow at different rates. Society members also collect information on sightings of leatherback turtles. These are only some of the projects being carried out under the society's auspices.

From small beginnings the society has developed into an organization that provides an effective forum for fishermen, scientists, and others with an interest in the long-term stewardship of marine resources to meet and deliberate in a collegial and nonconfrontational manner. It has built bridges between people who should have been collaborating from the outset and who now face the challenges of the future as partners.

The most important attribute of the society is that it provides a nonconfrontational forum in which fishermen, scientists, and other professionals can discuss and debate issues of common interest. Over time it has evolved into an organization that responds to the input of all members by facilitating goal-oriented research of interest to both parties and prerequisite to developing a long-term sustainable fishing industry on the east coast of Canada. The society has and continues to struggle to maintain an adequate funding base with which to further its objectives but demonstrates a tenacity derived from its members, who recognize its long-term importance.

The formal structure of the society consists of a number of elected unsalaried officers who act as a decision-making executive. The executive consists of a president, vice president, secretary, treasurer, chairs of the scientific program committee, communications committee, and a number of directors at large whose purpose is to survey the opinion of the members as to societal activities and projects. The scientific program committee was established to help fisherman and scientist members formulate, articulate, and design projects of mutual benefit to all members and consistent with the objectives of the society. The communications committee was in turn established to ensure that information from the general operations of the society and, more particularly, on the results of society (and related) projects was communicated to the membership. There are four main vehicles for communication. The first is a quarterly newsletter (*Hook, Line and Thinker*) distributed to all members and to an additional 1,000 subscribers in North America and abroad. The second is the Science Forum, held each year in conjunction with the annual general meeting of the society. The Science Forum now has an array of scientific presentations based either on original research carried out by society members or in areas of general interest to society members. It has attracted international participation, and this year's forum included a substantial poster session that consisted of short communications based on society research. Third, there is a Web site that contains detailed information on the society and its activities. Finally, society staff annually conduct a series of community meetings in which the results of research projects of interest to the participating communities are presented and discussed.

From the outset it was the objective of the society to become financially independent. This has not been achieved, and the most significant challenge for the long-term viability of the society remains funding. At present there is one main source of funding in the form of a management fee, charged to the government of Canada, for administering an annual longline survey of the eastern Scotian shelf. This survey is funded by the federal government to a total of approximately $300,000 per year. The government and the society have entered into a joint project agreement that clearly states each partner's responsibilities and the terms of the agreement. Other sources of funding for the society include contract work for the federal government, universities, and the fishing industry, as well as funds from a federally funded science and technology youth internship program. There are also donations from provincial governments, private-sector companies, fishing industry organizations, and donations from members.

These revenues have been sufficient to employ a manager for the administration of society business. The major business activities include: (1) administration of the annual longline survey of the eastern Scotian shelf, including vessel selection, ensuring compliance with survey protocols, payment to participants, and involvement in data analysis; (2) administration of a federally funded Science and Technology Student Youth Internship Program that provides opportunities for young people (from especially rural communities) to gain experience working in fisheries science projects; (3) searching out sources of funding to support the objectives of the society both in the form of research funds and operating funds; (4) producing a quarterly newsletter; (5) organizing regular meetings of the scientific and communications committees; (6) organizing the annual general meeting including the Science Forum (see above); and (7) managing the day-to-day operations of the society, including personnel management.

At present the society remains heavily dependent on government funding to remain viable. This is not an ideal condition in that the operation of the society is essentially at the mercy of government funding, with its inherent instabilities. This concern aside, the structure, function, and objectives of the society provide an effective linkage between fishermen and scientists that has aided in the communication between these communities and, as a result, increased the acceptability of resource assessments conducted by government scientists. It has also provided a forum for joint research aimed at furthering the objectives (long-term sustainability of the fisheries) of the society. Most recently the society was approached by the fishermen and scientists from the west coast of Canada asking it to consider starting a west coast chapter. Society representatives traveled to the west coast and met with a number of organizations there. The process has now become a more formal feasibility study to set up the west coast chapter of the FSRS.

It would appear that the society's objects and structure are therefore not dependent on the conditions and culture of eastern Canada and thus might be a model that the National Marine Fisheries Service (NMFS) could explore as a regional model in the United States. The most significant improvement that could be made if this model were to be explored by NMFS would be to establish a stable funding source.

BRITISH COLUMBIA SABLEFISH

The Canadian Sablefish Association (CSA) is an industry group composed of all license holders in the sablefish fishery in British Columbia. This is an individual transferable quota (ITQ) fishery, in which the average catch over the last 20 years has been about 4,000 tons per year with a market value of about $Can 20 million per year. This was one of the first Canadian ITQ fisheries and is one of the most profitable fisheries on the west coast.

Since 1991 the CSA (and its predecessor, the British Columbia Black Cod Fishermen's Association) has been actively involved in cooperative research with the Canadian government agency, the DFO. The cooperation is governed by a formal comanagement agreement that specifies the rights and responsibilities of all parties. A series of scientific and management committees are specified, with joint government industry participation on all committees. All decisions are reached by consensus, but the government, in the form of the Minister of Fisheries, reserves final authority over any conservation decisions and has traditionally set the actual harvest level each year.

The primary data collection methods used in the sablefish management are: (1) an annual tagging program, in which 20,000–30,000 sablefish are tagged with visible tags, and an associated tag recovery-reward system; (2) an annual pot survey to provide an index of abundance; (3) vessel logbooks to provide an index of abundance; and (4) biological sampling of fish captured at sea. In addition, there is an annual stock assessment process involving at present two DFO scientists and two consulting scientists employed by the CSA. In addition to these data collection and analysis projects, CSA has been intensively involved in the design and testing of escape rings, to allow smaller sablefish to escape the pots, and an entrance tunnel design to try to prevent large females from entering.

All of these projects are done cooperatively, with all being funded by the CSA and primarily operated by CSA. One government scientist is usually onboard for the survey and tagging programs. Most data collection and preliminary analysis are conducted by consultants employed by CSA, and CSA also funds one of the government scientist positions that maintains the databases on DFO computers. The CSA-DFO level of cooperation is much more intensive than any that have evolved so far within the United States.

BRITISH COLUMBIA GROUNDFISH TRAWL FISHERY

The British Columbia groundfish trawl fishery is an ITQ fishery with approximately 120 active vessels. Two major changes to the fishery took place in the 1990s. The first was the requirement of 100 percent observer coverage (paid for by vessel operators), mandated in 1993. The fishery is required to retain all catch, and any catch over vessel trip limits is surrendered on landing, with proceeds going into a research fund administered by the fishing industry. This provided what was then an economically struggling industry the economic resources to begin participation in the fishery management process. The second major change was a move from trip limits to ITQs in 1996, which led to substantial economic rationalization and an increase in profitability for remaining operators.

The groundfish trawl industry, through its organization the Canadian Groundfish Conservation Society (CGCS), now funds two primary cooperative research activities. It has instituted an industry-funded and -operated deepwater trawl survey to provide data on the important deepwater species. Industry consultants design and operate the survey, which takes place with government scientists onboard. The CGCS consultant also works within the DFO stock assessment system, including the preparation of annual stock assessments and participation in the stock assessment review process.

The initial impetus to industry involvement in cooperative research was the funding available from the revenue derived from landings above and beyond trip limits. Before the fishery was profitable, there was little industry willingness for self-funding research.

A final characteristic of the British Columbia groundfish trawl fishery is that there are no ongoing government-funded fishery-independent surveys. Over the last 20 years there have been a number of surveys, but there has been no consistently funded program similar to those in most of the eastern Canadian and U.S. fisheries.

COOPERATIVE RESEARCH IN NEW ZEALAND

Fisheries research in New Zealand needs to be considered in the context of three characteristics of the management regimen currently being used: (1) the system of ITQs and the rights and obligations that these quotas confer on their owners; (2) the requirement on the Minister of Fisheries (MFish) under the 1983 and 1996 fisheries acts to maintain fish stocks at or above the maximum sustainable yield level and requirements under these acts and other acts that cover protected marine mammal and seabird species and also contain substantial ecosystem sustainability requirements; and (3) the recovery of all directly attributable costs from the commercial fishing industry; including all research, management, administrative, and enforcement costs.

There are several categories of fisheries-related research that are undertaken in New Zealand:

- "Required services" are research projects that are undertaken in support of the MFish's obligation to maintain fish stocks at sustainable levels. These services are selected through a research planning process, but the final determination is by the officials of the MFish. These required services include research to address ecosystem sustainability issues as well as fish stock sustainability.
- Elective research is undertaken by the fishing industry or the Crown to address issues of importance that are not being addressed through the required services described above.
- General marine science research is funded directly by the New Zealand government. This type of research has little overlap with fisheries issues and will not be discussed further.

MFish has little science capacity of its own. There are approximately six full-time staff who operate the research planning and stock assessment reporting processes as well as writing and evaluating the research tenders. They also interpret scientific advice to the MFish. The large cadre of research science staff who were previously government employees have been devolved (in 1995) to a Crown-owned corporation, which is run like a business and is established by legislation requiring the corporation to make a profit. All approved research projects are written as tenders and are offered to the "best" bidder (as determined by a relatively transparent evaluation procedure). In practice, the former government fisheries research organization (now called the National Institute of Water and Atmospheric Research, or NIWA) wins about 95 percent of the contracts.

There is an annual cycle that defines the New Zealand fisheries research process. The following description attempts to summarize this cycle, but it should be noted that the dates used to characterize each step of the process are only indicative of the general time period involved.

Research planning occurs from August to October. This is where the various potential required services projects are discussed and evaluated. All "stakeholders" are free to participate, including the scientific staff from NIWA. These individuals were excluded from this step early on in the development of this process, presumably because it was thought they were in a position of conflict of interest. But this policy was dropped as it became clear that much of the research expertise was being excluded from the discussions. The final list of potential projects is generally arrived at through a consensus procedure, but MFish always reserves the option of making the final decision on what research projects go forward.

The release of the "white paper" occurs in December and January. The white paper is a listing of the specifications for the majority of the approved projects, although other projects may be tendered at different times of the year. Intermediate and final deliverables are specified for every research project, which imposes a high level of accountability on the completion of the project, given that MFish does not pay until the deliverables have been provided. Also notable is that each quota holder is proportionally assessed the costs associated with each project, based on a determination of how each fish stock benefits from the research and the quota holdings.

From January to April, fishery assessment working groups are convened. This is one of the most important functions of the annual research cycle. Beginning in late January, working groups associated with single species or with species groupings are convened and the results of previously contracted research are presented. These working groups provide peer review of the work and, in the case of stock assessments, also provide consensus feedback for work in progress. Scientists and other fishing industry representatives have been active participants in this process since around 1990. They have also directly contributed to the assessment work, either by working on the assessments independently or, more recently, as full collaborators with NIWA scientists.

A fishery assessment plenary occurs from late April or early May. This meeting is the culmination of the working group process where the most important (or contentious) assessments are presented to a wider audience than attends the working groups. This provides a second opportunity to review the work. The final output of the working group process is a summary of the assessment information for each species being harvested in the New Zealand exclusive economic zone, which becomes the main input into the scientific advice given to the MFish.

Management advice is given to the MFish from late May to September. The Minister needs to make most of his management decisions in sufficient time that they can be implemented on October 1 in each year. The scientific information that is generated through the research process is integrated with other information to provide the Minister with the basis to make informed decisions. This process is long and involves a considerable amount of consultation with all affected stakeholders. However, it should be noted that the paramount concern in the decision-making process is the stock status for the fish stocks being harvested, and in recent years ecosystem sustainability is also becoming a consideration in the decision-making process. These sustainability obligations are established through legislation and are an important constraint on the government in making management decisions.

The involvement of the New Zealand fishing industry in the fisheries research process is large and complex, and the fishing industry contributes to the research process at every level.

As noted above, the fishing industry has employed scientists to participate at the fishery assessment working group level since the late 1980s. Assessments parallel to those performed by government scientists have been performed since 1990, particularly for hoki (which is the species taken in the largest tonnage). Recently, a cooperative model in which assessment research is done jointly by NIWA and fishing industry scientists has been evolving, particularly in the hoki and orange roughy assessment work, and it is expected that this will become a standard approach in future years.

The Rock Lobster Industry Council (RLIC) is in the unique position of being the provider of research services to the MFish, as it won a three-year contract to provide these services (NIWA was the other bidder). RLIC subcontracts the assessment research to three independent scientists, of whom only one works for NIWA. NIWA also has the subcontractual responsibility to provide sampling technicians for some fisheries. The rest of the research, including a substantial tagging program and a "logbook" program (self-monitoring sampling of the catch), is either administered directly by RLIC or is performed by one of several smaller commercial fishery stakeholder groups. The assessment research was independently reviewed in 2001, with no major criticisms.

The fishing industry has frequently commissioned and funded additional research to address issues that it feels are not being properly addressed through the main MFish process. The parallel and cooperative assessments described in a previous paragraph fall into this category. As well, various parts of the New Zealand fishing industry have not felt comfortable with some aspects of the government-sponsored research and, consequently, have encouraged the development of an independent science capacity in New Zealand. In particular, the orange roughy quota holders have commissioned independent acoustic surveys to complement those that were required by MFish. These surveys have been conducted by other research institutions from Australia and South Africa using commercial fishing vessels as research platforms. In addition, the orange roughy industry component is developing methods that make use of active fishing vessels to undertake acoustic surveys as part of their fishing operations.

The New Zealand rock lobster industry also operates a logbook and catch sampling program that now provides the basic data input to the lobster stock assessments. Regional fishing groups employ scientists to design and monitor their sampling program. After making a comparison to the government sampling program, stock assessment scientists concluded that the industry-based program was suitable and indeed more extensive than the government-operated program, and the latter was terminated. The industry-based program was initiated by a small group of fishermen in one part of New Zealand who felt that their fishery was considerably healthier than several others but that, due to the design of the government's sampling program, they could not show this. They funded their own program, which successfully demonstrated the health of their stocks, and this program then spread to regional groups throughout the entire country.

There are a number of sampling projects that have been undertaken directly by the New Zealand fishing industry to address their concerns about the nature of the catch in some fisheries. For instance, there was a strong belief within the hoki catching sector that the proportion of small hoki taken on the Chatham Rise was too high to maintain a sustainable fishery. But the level of observer coverage funded through the MFish process was too low to determine the composition of this catch accurately. An onboard self-monitoring program was designed by the New Zealand Seafood Industry Council (SeaFIC) science staff to address this issue, and this program has been operating on these deepwater vessels since mid-1999 (over 2,000 length samples representing nearly 250,000 length measurements were taken in 2000–2001). Onboard sampling has also been instituted in the orange roughy fleet, and onshore factory sampling for biological characteristics has also been independently commissioned by various components of the New Zealand fishing industry in snapper, orange roughy, and some pelagic fisheries.

In 1991, MFish and the fishing industry agreed to institute a program of experimental quota increases in return for additional monitoring work on some small inshore fisheries. This approach has been frequently used to address overcatch issues in bycatch fisheries, with intermittent success. However, the adoption of these programs has led in several instances to strong onboard self-monitoring programs, as well as improved attention to these fisheries (in comparison to what would have been in place without the program).

Several fisheries in New Zealand capture protected marine mammal and seabird species, and the existence of this interaction has led in several cases to substantial amounts of additional research being undertaken by industry. For instance, the bycatch of Hooker's (New Zealand) sea lion in the Auckland Islands squid fishery has generated a substantial amount of population modeling research on this mammal species. A wide range of fisheries scientists have worked on this problem on behalf of the squid industry, including some from NIWA. The interaction with this sea lion species has led to a great deal of industry-funded research to develop excluder devices that can minimize the effect of fishery-related mortalities on this population. Similarly, the bycatch of protected seabirds has led to a number of industry-sponsored research projects, including a project to develop an experimental design for seabird population studies and other projects to test the effectiveness of mitigation devices like tori lines and bait-delivering schemes.

The New Zealand fishing industry has demonstrated a strong commitment to cooperative fisheries research in support of the fishery resource on which they depend. Total expenditures on "required" fisheries services is currently on the order of about New Zealand $18M. The New Zealand fishing industry is probably spending on the order of New Zealand $3M (this is an estimate) on additional fisheries research, including maintaining a significant science presence at SeaFIC. The distinction between "cooperative" and "government" research is becoming increasingly blurred in New Zealand, given that the RLIC and the challenger scallop fishery, which are both industry stakeholder groups, are providers of rock lobster and scallop scientific information for MFish while NIWA is performing significant amounts of work directly for the fishing industry.

MFish and the Environment Minister retain the ultimate responsibility for conservation of the aquatic resources and can unilaterally stop any fishing action if they believe it poses a conservation threat. MFish in late 2002 closed the setnet fishery on the west coast of the North Island because of bycatch of Hector's dolphin. While New Zealand has not yet found an operational definition of ecosystem management, both ministers can take whatever action they feel is required to protect ecosystems, and this has included setting aside a number of large areas that are now closed to fishing.

SUMMARY OF INTERNATIONAL EXPERIENCE

This chapter has presented illustrations of cooperative research in eastern Canada, western Canada, and New Zealand. The eastern Canadian examples are largely government funded and are designed to provide studies additional to the ongoing government research programs. In western Canada the activities have been industry funded, and the data collection programs provide the core of the scientific activity for the fisheries. The differences between western and eastern Canada are numerous. In the west, the user groups are small, have individual quotas of the catch, and are profitable. In much of eastern Canada the individual quotas are not guaranteed and the fisheries are generally overcapitalized.

On the east coast of Canada, cooperative research mainly takes the form of abundance surveys, some funded by the government and others funded by fishermen having property rights to the fish they harvest. A review of the government-funded surveys concluded that standardized data collection protocols with clearly defined roles for each partner, coupled with quality assurance through onboard observers, led to more successful partnerships. Information collected by cooperative surveys both augments existing information and provides new insights into the functioning of the harvested populations. Cooperative surveys have also led to the conduct of ancillary research of mutual interest to the partners. These partnerships are likely to become part of doing business in the east coast fisheries of the future. For those examples where the survey is self-funded and where the industry partners had a sense of ownership (of both the resource and the survey), there was willingness to make a long-term commitment. This long-term commitment may have resulted from necessity to have a better estimate of resource abundance, vested interest, and the real possibility of economic losses in the absence of the survey. The FSRS may provide a model to facilitate the development and implementation of cooperative research.

New Zealand provides a substantial contrast, where the fishermen have rights similar to those in western Canada, but there is also the legal requirement of industry funding of all management expenses. This provides enormous incentive for the fishing industry to develop its own research programs that are as cost effective as possible. The fact that countries with quota share–based management and cost recovery conduct significant cooperative research suggests that there are economic imperatives due to lower costs, higher quantity and quality of research and management outputs, and/or enhanced value of resource rents and marketable quotas. The western Canadian examples are perhaps the most surprising, where these two industry groups have devoted substantial research programs at their own expense without a legal mandate to pay for the research costs. In both examples the industry felt it would end up losing potential catch without better scientific research—a motivation also seen in many U.S. fisheries, but in those where the fishing industry was able to either obtain funds from government or set aside catch to fund the research.

4

Setting Research Priorities

IDENTIFYING COOPERATIVE FISHERY
RESEARCH NEEDS AND OPPORTUNITIES

Regardless of whether a fishery research program is labeled as fundamentally cooperative, collaborative, noncooperative, directed, or traditional, effective research programs must identify research needs and priorities consistent with legal requirements, management objectives, and budget and resource constraints. Identifying and prioritizing research needs in fisheries, however, can be a daunting task given the (1) challenging legal and regulatory environment, (2) multiple and potentially conflicting management objectives, (3) significant scientific uncertainties, (4) numerous stakeholder groups with alternative agendas, and (5) limits on fiscal and human resources.

Under the reauthorization of the Magnuson-Steven Fishery Conservation and Management Act (MSFCMA) as amended by the Sustainable Fisheries Act in 1996, Section 404 (Fisheries Research) requires the Secretary of Commerce to develop a strategic fisheries research plan (NMFS, 2001). The act stipulates that the plan provide a role for commercial fisheries in research areas described within the plan, including involvement in field testing. The plan requires that a comprehensive program contain the following (priority) areas of research:

- Research supporting fishery conservation and management, including but not limited to biological research, abundance, trends, life history of stocks of fish, interdependence of fisheries or stocks and the ecosystem, identification of essential fish habitat, impact of pollution, and impact of wetland and estuarine degradation

   • Conservation engineering research, including the study of fish behavior, developing and testing new gear technology and fishing techniques to minimize bycatch and any adverse effects on essential fish habitat, and the promotion of efficient harvest of target species

   • Research on the fisheries, including the social, cultural, and economic relationships among fishing vessel owners, crews, processors, labor, seafood markets, and fishing communities

   • Research and development of a fishery information base and an information management system

The emphasis on constituent involvement in cooperative research is reemphasized in the National Oceanic and Atmospheric Administration (NOAA) Fisheries Strategic Plan (NOAA Fisheries, 2002), which states:

> To the extent practicable, we (NOAA Fisheries) will charter fishing vessels to participate in research projects, invite constituents to participate aboard NOAA research vessels during resource surveys, encourage frequent contact and cooperation between scientists and constituents, and incorporate scientifically valid observations by fishers and others into fish stock assessments and other analyses related to living marine resources and their habitat.

Goal 5 of this plan states that National Marine Fisheries Service (NMFS) will improve the effectiveness of external partnerships with fishers, managers, scientists, conservationists, and other interested parties by:

   • Promoting a cooperative network of partners in the coordination of fisheries research
   • Developing infrastructure for long-term continuous working relationships with partners to address fisheries research issues
   • Sponsoring symposia and conferences for partners to exchange information and identify major research initiatives
   • Soliciting partners' views on fisheries research needs

Developing a mechanism and infrastructure to develop, prioritize, and coordinate cooperative research, however, can be a daunting task given the overall complexities of the U.S. fisheries science and management system and issues specific to each region and fishery. For example, in 1998 a conference was held in Portland, Oregon, with an explicit objective of developing a research infrastructure for prioritizing and coordinating West Coast cooperative groundfish research (Fisher et al., 1999). Although many good ideas and proposals were developed, none of the participating groups, including the Pacific Fisheries Management Council (PFMC), Pacific States Marine Fish Commission, Northwest Fisheries Science Center (NWFSC), state agencies, environmental organizations, industry groups, Sea Grant, and universities were willing to take leadership and champion the development of a coordinating mechanism at that time. Although some of the research ideas were valuable to NWFSC in developing the groundfish research plan, no coordinating infrastructure has been developed on the West Coast that is consistent with NOAA Fisheries Strategic Plan Goal 5.

A recent report by the National Research Council (NRC, 2002) emphasizes five areas of science that are not adequately addressed by NMFS and should be considered high priority: (1) research to meet the legal mandates of the Marine Mammal Protection Act and Endangered Species Act; (2) collecting and analyzing spatial data; (3) supporting ecosystem science and related models; (4) developing new techniques to link biological, social, and economic data; and (5) linking market and nonmarket values with management scenarios. The report also states that NMFS should facilitate greater cooperation between scientists and stakeholders (including fishery participants) to improve the quality and efficiency of data collection in these and other areas of science and create a shared sense of confidence in what the data indicate.

The MSFCMA, NMFS strategic plans, and NRC reports highlight and prioritize broad areas of fisheries research. These documents also emphasize the importance and potential contribution of industry and stakeholder cooperation for improving science and management. Although these documents explicitly note cooperative research that involves vessels, gear, or fisherman's knowledge, they do not prioritize any particular research area for cooperation; nor do they discuss research areas that should not be prioritized or may be inappropriate for cooperation. The implicit but unstated assumption is that many types of research may be improved through cooperation in the science process. This can include stock assessment and monitoring, gear-related research on impacts to the habitat and ecosystem, and social and economic research requiring sensitive individual- and firm-level data.

An additional but equally important issue is determining what degree of cooperation will maximize benefits: cooperation in only one element of the research and science process or comprehensive engagement ("collaboration") in most or all elements of the research process (ideas, hypotheses, proposals, funding, design, conduct, analysis, review, publication, communication). The significant regional differences in fisheries, ecosystems, management and scientific institutions, and fiscal resources suggest there is no consistent answer for determining which cooperative research projects should be prioritized. However, on the basis of recent experiences in cooperative research, there are some guiding principles and criteria that may be useful for prioritizing cooperative research and estimating the degree of cooperative engagement that will generate significant, positive benefits for science, management, and society.

DEVELOPING CRITERIA FOR PRIORITIZING COOPERATIVE FISHERIES RESEARCH

Any process developed for prioritizing cooperative fisheries research must be efficient, open, transparent, and fair. In addition, the following issues should be considered:

- The expected gain in scientific and management benefits
- The types and degree of cooperative engagement that will maximize fishery science and management benefits

In some cases only a small degree of engagement may be appropriate and necessary to generate substantial research and management benefits. For relatively minor fisheries research (i.e., an expected low payoff), the transaction costs to engage fishermen and other constituents may exceed possible science and management benefits

- The expected research costs, including opportunity costs of employing fiscal and human resources
- The expected time stream of net benefits (e.g., short-term versus long-term net payoffs)
- The efficiency, openness (transparency), and fairness of the prioritization process
- The process can be led by a single group or a committee, but the process and results cannot be owned by any given organization.

The process must include all relevant constituents and partners in selecting and prioritizing cooperative science and management.

- The objectives of cooperative research

Fishery management is plagued by numerous objectives that are vaguely defined, conflicting, or unquantified. This makes it difficult to develop criteria for prioritizing cooperative research. It is critical that scientists and other parties engage fishery managers in strategic discussions for establishing quantifiable objectives that can be used for prioritizing and evaluating cooperative research (NRC, 2002).

In addition, scientists, managers, industry, and other constituents need to collaboratively evaluate potential benefits and costs over time in order to develop consensus priorities for cooperative research. For example, a study by Harms and Sylvia (1999) demonstrated that while West Coast groundfish fishermen and scientists had significantly different views on the potential benefits of alternative science-related cooperative projects, they had similar views on the relative costs and potential cost effectiveness of these projects.

Innovation and competition are key elements driving an efficient scientific process. A fair but incentive-based process for awarding cooperative fisheries research funds is critical for achieving science and management objectives. Criteria must be selected to ensure that innovation, creativity, and rational competition are maintained while emphasizing equity and fairness. Government administration and industry can collaborate as teams, but there may be direct competition between teams with different ideas, hypotheses, and research methods.

Evaluating successful and unsuccessful cooperative research projects can be useful for prioritizing and estimating the expected benefits and costs of alternative cooperative research projects and their design and level of engagement. The elements of successful projects include (1) substantial incentives and benefits to research partners; (2) rigorous coengagement in most elements of the scientific process; (3) complementary skills and abilities; (4) honesty, trust, and mutual respect; and (5) adequate financial, administrative, and scientific support. Successful and popular cooperative research also tends to cluster around projects where fishermen's vessels, gear, and expertise can be readily employed (e.g., fisheries monitoring, bycatch studies, logbooks), and research results may substantially change assessments and regulations to provide short- and long-term economic benefits to fishermen. Although cooperative projects have not tended to focus on long-term environmental studies on ecosystems and fishery habitats, this may reflect institutional disincentives rather than lack of interest on the part of fishermen in the long-term health and productivity of the marine environment.

The National Academy of Public Administration (NAPA) noted that NOAA Fisheries (NMFS) is criticized for its lack of openness in establishing science programs. NAPA recommended that NMFS jointly develop and administer its research programs with key constituents (NAPA, 2002). NAPA also noted that there is no standardized process for selecting and prioritizing cooperative research for federally managed fisheries. Each region, science center, and council has alternative approaches for developing and prioritizing cooperative research. For example, the NWFSC has developed a comprehensive and prioritized groundfish research plan (NWFSC, 2000) that uses ideas, discussions, and recommendations from various forums, individuals, and constituent groups. The plan describes the general need for industry and constituent involvement but provides no details or strategies. The PFMC's Science and Statistical Team develops a biannual prioritized Research and Data Needs document. These needs, however, are developed without formal coordination with the NWFSC. The Research and Data Needs document also does not discuss or prioritize cooperative research projects. In contrast, the New England Fisheries Management Council has established the Research Steering Committee composed of scientists, managers, and industry to develop prioritized cooperative research projects that integrate science and management needs. Projects are supported through congressional funding to NMFS and specifically the Northeast Fisheries Science Center, which are targeted for cooperative research.

ALTERNATIVE PROCESSES FOR PRIORITIZING AND COORDINATING COOPERATIVE FISHERY RESEARCH

Depending on politics, institutions, and funding, there are a number of possible policy models or "policy infrastructure" scenarios that could be used to prioritize and coordinate cooperative research while addressing NOAA Fisheries Strategic Plan Goal 5. The following descriptions briefly summarize six examples.

Scenario 1: Status Quo Cooperative Research Coordination

The following are characteristics of current coordination of U.S. cooperative research:

- no standard approach for planning and prioritizing cooperative research across regions
- no structure to coordinate cooperative research and management across constituents, management agencies, and science groups, including centers, councils, and universities
- uneven distribution of earmarked cooperative research funds across regions and among science centers
- schisms between industry, NMFS fisheries science centers, and other agencies over prioritization and allocation of earmarked dollars for "cooperative" versus "noncooperative" research
- significant input from Congress over support and design of "cooperative" research
- unstable funding for NMFS fisheries science center research budget line items and earmarks for cooperative research

This complicates research planning and potentially reduces effectiveness of all cooperative research programs.

Scenario 2: NMFS-Based Cooperative Research Coordination

An alternative approach would be to directly provide NMFS with the administrative authority to coordinate cooperative research. This would be characterized by:

- NMFS fisheries science centers having a lead role in coordinating and prioritizing federal fisheries research within each region, including directing funds earmarked for cooperative research
- national headquarters providing oversight and/or standard approaches for coordinating or prioritizing
- NMFS fisheries science centers electing to form advisory groups for science and management, and constituency groups to assist in developing priority research and recommend approaches for integrating cooperative effects for all priority research areas
- potential for rifts between industry, other entities, and NMFS fisheries science centers over prioritization and allocation of earmarked dollars
- Unstable funding for NMFS fisheries science center research budget line items and earmarks for cooperative research, complicating research

planning and potentially reducing effectiveness of all cooperative research programs.

Scenario 3: Fishery Management Council-Based Cooperative Research Coordination

If the fishery management councils (FMCs) were to take the lead we could see:

- FMCs lead the effort to select, prioritize, and coordinate cooperative fisheries research funded with federal funds within each region, including funds earmarked for cooperative research
- each FMC establish a cooperative research committee composed of a broad cross section of federal, state, academic, and nongovernment scientists, industry representatives, and fishery managers
- FMCs provide cooperative research committees with broad discretion to select and prioritize cooperative research areas and programs to achieve fishery management goals

Cooperative research committees would be expected to evaluate the potential engagement opportunities for industry, nongovernmental organizations (NGOs), and other private and public groups for all federal fisheries science. The committees would be expected to develop formal coordination plans with the NMFS fisheries science centers.

- potential for rifts between NMFS fisheries science centers, environmental organizations, and FMCs over committee recommendations
- potential for continued pressure on Congress to earmark funds for "cooperative" research, leading to concern about NMFS fisheries science center funding
- unstable funding for cooperative research, complicating research planning and potentially reducing effectiveness of all cooperative research programs

Scenario 4: Industry-Based Research Coordination

Fishing industry organizations could lead the effort to select, prioritize, and coordinate industry-funded fisheries research through the following:

• industry organizations impose voluntary tax or, conversely, government requires industry groups to pay for research and management costs as mandated by "cost recovery" programs (cost recovery programs are usually associated with quota share-based fishery management systems, e.g., individual fishing quotas [IFQs])

• industry groups use a coordinating committee to reach consensus with scientists and government in selecting, prioritizing, and conducting research (e.g., West Coast Canadian Groundfish IFQ Program; Turris, 1999)

• NMFS fisheries science centers not directly responsible for cooperative fishery research attributable to fishery management; however, they might contract with industry groups in cooperative ventures

• potential rifts between industry, NMFS fisheries science centers, and FMCs over committee recommendations and study outcomes

• unstable funding as a result of variations in revenues available for taxation or cost recoveries, complicating research planning

Scenario 5: Neutral Third-Party Research Coordination

Other organizations have missions consistent with leadership roles in prioritizing and coordinating fisheries research. For example, national or state Sea Grant organizations, interstate marine fisheries commissions, or regional fisheries foundations could play vital roles in prioritizing and coordinating federal fisheries research. These organizations are perceived as neutral third parties relative to the FMCs, industry groups, environmental organizations, and NMFS fisheries science centers. A number of New England Sea Grant programs play a partial role in encouraging and funding cooperative fisheries research through the Northeast Consortium. Such neutral third parties could also play a pivotal role in helping NMFS fisheries science centers implement comprehensive cooperative fisheries research. However, it should be noted that there still is potential for rifts between NMFS fisheries science centers, environmental organizations, and FMCs under this scenario. This is because the decisions would still be made by a single organization, which may be perceived as not representing all constituents.

Scenario 6: Regional Research Boards

An alternative to either the status quo or nonbinding coordinating policy processes is to establish regional research boards with the authority to prioritize, coordinate, and evaluate the use of funds allocated to cooperative fisheries research projects (either as earmarks or line items). The regional boards could also assist NMFS in identifying dedicated research (research not currently conducted as cooperative research) that might be good candidates for cooperative research.

The regional research boards would be nonprofit organizations that would be funded by the federal government but would also have the ability to receive private funding from other sources and to support multiyear projects. At the regional level, this advice would be provided to the NMFS regional office, regional NMFS fisheries science center, and the appropriate FMC. At the national level, the advice would be provided to the NMFS national office from all of the regional boards through a national steering committee consisting of the chairs of the regional boards. The regional research boards might have the following responsibilities and structure:

- The regional research boards could be a nonprofit organization funded by the federal government.
- The regional research boards could be administratively independent or could operate under the umbrella of the regional FMCs.
- The regional research boards would be composed of a broad range of members, including leading scientists, and other constituents, including representatives from the regional NMFS fisheries science center, state government, industry, academia, regional FMCs, and NGOs. Nonfederal board members could receive compensation. All board members would be expected to participate in a training program commensurate with their specific duties and responsibilities.
- The primary function of the regional research boards would be to prioritize, coordinate, and evaluate all federally funded cooperative fisheries research within each region, consistent with the objectives of the MSFCMA. The boards would be expected to develop consistent and objective criteria for selecting and prioritizing cooperative fisheries research areas, projects, and programs. They would work closely with the regional NMFS fisheries science center and the regional FMC(s).
- A secondary function would be to evaluate the potential types and levels of engagement of industry, NGOs, state governments, and other con-

stituents in order to improve the success, innovativeness, and cost effectiveness of cooperative fisheries research within the region. The board would be expected to establish rigorous, incentive-based scientific cooperative research program protocols to meet these objectives.

• The regional research boards would be expected to conduct symposiums and foster other methods of communication to engage all constituent groups and scientists in sharing research ideas and information.

• The regional research boards could also evaluate all federally funded research in the region for its potential as cooperative research.

• The regional research boards would receive funding for staff to support the activities of the regional research boards.

5

Funding, Legal Issues, and Scientific Rigor

FUNDING SOURCES AND DISTRIBUTION

Cooperative research has been funded by a broad range of methods, including direct congressional mandates, congressional line items, funds directed to cooperative research from National Marine Fisheries Service (NMFS) headquarters, funds from NMFS fisheries science centers or regional offices, industry set-aside funds, import duties, landing taxes, direct industry contributions, and funding from nongovernmental organizations (NGOs) and foundations. These sources are mutually compatible, and indeed, we believe that many if not all of these sources of funds will continue to support cooperative research in the future. However, many of the funding sources deserve comment, particularly with respect to the recommendations made below (see Table 5-1).

Over the last several years, Congress has taken a much more active role in determining the allocation of funds to individual cooperative research programs and projects. Some congressional funds now appear as line items for specific projects or under a general heading of "cooperative research" within a certain region of the NMFS budget. Alternatively, Congress, NMFS, and other agencies like the U.S. Fish and Wildlife Service may provide direct funding for a regional entity such as the Northeast Consortium, which then decides how to allocate funds across competing projects. An important feature of both of these methods is that specific funds are set

TABLE 5-1 Summary of Current NMFS Grants and Programs Available for Cooperative Research

Program	FY 2002 (in millions)[a]	FY 2003 (in millions)	Competitive Process	NMFS Partner Required	Industry Partner Required?
Alaska Fisheries Development Foundation	$0.75	0	Yes	No	Yes
Alaska Fisheries Information Network	$3.2	$3.2	Yes	No	Yes
Gulf and South Atlantic Fisheries Development Foundation	$0.4	0	Yes	No	Yes
Gulf of Mexico Fisheries Information Network	$3.5	$3.5	Yes	No	Yes
GoMex Consortium	$2.75	$1.25	Yes	No	Yes
Interjurisdictional fisheries grants	$2.59	$2.59	Yes	No	No
Marine Fisheries Initiative (MARFIN)	$3.5	$3.5	Yes	No	Yes
Northeast Consortium	$5.0	0	Yes	No	Yes
Pacific Coast Information Network (PACFIN)	$3.0	$3.0	Yes	No	Yes
Recreational Fisheries Information Network (RECFIN)	$3.7	$3.7	Yes	No	Yes
Saltonstall-Kennedy Grants Program	?	$4.1	Yes	No	No
Sea Grant	$62.4	$57	Yes	No	No

[a] These are in addition to line items specifically appropriated for cooperative research. They do not include grant, loan, and development programs whose purpose is business development rather than research.
SOURCE: NOAA (2002).

aside for cooperative research, without an evaluation of the economic or scientific merits of cooperative versus dedicated research.

NMFS also funds cooperative research from within its own budget, both at the regional office and the fisheries science centers levels. In this case, it is NMFS that determines which projects to fund, which projects to do cooperatively, and what to do solely with NMFS staff. An advantage of this approach is that it provides for a more deliberate consideration by government of the economic and scientific efficiencies of different levels of cooperative work, but it also has the disadvantage of being funded within the constraints of an organization attempting to maintain its own institutional infrastructure.

Some funds for cooperative research come from specific set-asides and import taxes. The Saltonstall-Kennedy Grants Program is an example of such programs, and there now exist project decision and administrative frameworks for allocation of such funds. The amount of funds available is determined by legislation, but it would certainly be possible to imagine using landing taxes as a primary funding mechanism for most data collection. In New Zealand a regional fishery management board determines the appropriate budget for specific fisheries and sets landing taxes based on the estimated costs. Advantages of this system are that the amount of funds spent on a fishery is determined by the management needs, a "user pay" cost recovery management system with associated stakeholder involvement in the decision making, and potentially less political influence in the funding process.

The committee heard several examples of direct industry funding of projects, and this is certainly a trend to be encouraged. It seems that such funding will occur when industry sees an immediate benefit to its economic interest, and provided such programs meet the scientific criteria of the decision-making process, the committee finds no problem with such projects. NGOs and foundations have been less involved in funding cooperative research, but some funding has occurred and, as with industry funding of projects, such funding should be encouraged.

LEGAL ASPECTS OF COOPERATIVE RESEARCH

There are a number of legal issues that participants in cooperative research projects need to consider. These issues include vessel licensing and inspection requirements, fishery permits, charter agreements and contracts, insurance, enforcement, and confidentiality and ownership of research data and information. Failure of participants involved in cooperative research to consider and address these and other legal issues can result in project delays, increased project costs, or project failure. Some of these important legal issues are discussed below.

U.S. Coast Guard Licensing and Inspection Requirements

A critical element of cooperative research is the chartering of fishing vessels as research vessels by NMFS. NMFS, fishermen, and other participants involved in cooperative research need to be aware of and meet all U.S. Coast Guard licensing and inspection requirements as they relate to the performance of each research project using fishing vessels. A brief description follows below.

The U.S. Coast Guard has for decades had safety regulations in place covering oceanographic research vessels and passenger-carrying vessels and more recently has established safety regulations for fishing vessels. These regulations range from the designation of vessels for particular uses, to the inspection and certification of vessels for designated uses, to the licensing of operators and crew aboard those vessels as required.

Vessels under 100 tons engaged in recreational fishing but carrying passengers for hire range from small motor vessels chartered for a fee to carry one recreational fisherman to large motor vessels taking 100 or more recreational fishermen to fishing grounds for a simple fee. When carrying fewer than six passengers, these vessels require an operator licensed by the U.S. Coast Guard but do not have to be inspected. Vessels carrying more than six passengers require both a licensed master and certification. Vessels engaged in fishing and so designated are not required to have such certification, but if the vessel is over 200 gross tons (GT), it must be operated by a master licensed by the U.S. Coast Guard. Oceanographic research includes the taking of biological samples from the sea, including fishery resources. Fisheries research, a type of oceanographic research, is interpreted to include environmental and fishery resource surveys and experimental fishing trials. Vessels engaged exclusively in oceanographic research are not required to have certification if under 300 GT, and the operators are not required to be licensed if the vessel is under 200 GT. An uninspected motor vessel of under 300 GT (such as a designated fishing vessel) operating exclusively as an oceanographic research vessel when chartered for scientific research is addressed specifically in U.S. Coast Guard policy. For long-term research vessel charters, these vessels may request a letter of designation from the local Coast Guard officer in charge, marine inspection (OCMI). After determining the vessel is employed exclusively in oceanographic research, the OCMI will issue a letter of designation for a period of two years. If the vessel receives this letter, the scientific personnel onboard are not considered crew members. An uninspected motor vessel of under 300 GT that does not possess a letter of designation may also engage in oceanographic research operations provided that such use does not violate applicable manning and inspection requirements. However, on these "undesignated" oceanographic research vessels, scientific personnel who serve in any capacity on vessels of at least 100 GT require merchant mariner's documents (MMDs) as members of the crew. In addition, the carriage of students is considered carriage of passengers, depending on the size of the vessel and number of passengers onboard. This compels Coast Guard certification and licensing of the vessel's master.

In summary, fishing vessels under 100 tons can be chartered for oceanographic research, and an operator or captain licensed by the U.S Coast Guard is not required as the scientific party is considered part of the crew. Fishing vessels over 100 tons can also be chartered to conduct oceanographic research; however, unless the vessel is issued a letter of designation, the entire crew, including the scientific party, must carry MMDs.

Fishery Permits

The need for fishery permits varies with the scope and type of the cooperative research projects. In the case where a fishing vessel is carrying a scientific sea sampler or observer that is collecting data under normal legal fishing operations, no special permits other than those required to legally conduct normal fishing operations within restrictions of the relevant fishery management plans are required. If a fishing vessel has been chartered or will be otherwise engaged in scientific research that will not require the sampling or taking of fishery resources, no fishery permits are required. If a fishing vessel has been chartered or will otherwise be engaged in scientific research that will require the sampling or taking of fishery resources with minimal impact on the stock or habitat, the leader of the scientific research project must submit a research plan to the NMFS regional director at least 60 days prior to the proposed activity, and after review the NMFS regional office may issue a letter of acknowledgment (LOA) to the project and the specified fishing vessel, authorizing the activity and specifying reporting procedures. If a fishing vessel is chartered or otherwise engaged in the sampling or taking of fishery resources where the catches may be retained and sold to offset the cost of the research, or the catches are potentially large enough to impact the goals of a fishery management plan, an exempted fishing permit (EFP) is required. To obtain an EFP, the research project leaders must submit a research plan and application to the NMFS regional director at least 60 days prior to the proposed activity. After review, the NMFS regional office may issue an EFP specifying the conditions of the allowed research activity and the reporting procedures.

Research activities that require the sampling or taking of fishery resources have been problematic for NMFS, scientific researchers, and the operators of fishing vessels. One example is the participation of a fishing vessel in a scientific survey where tow duration is short, liners are used to retain small animals, the catches are returned to the sea, and enumeration and measurement are clearly covered by the LOA. For some EFP applications, an environmental assessment (EA) may also be required because the environmental impact of the proposed fishing activity is believed to the substantial. The preparation of an EA requires considerable effort and expertise, and the criteria for when an EA is required vary from region to region.

In summary, there can be considerable confusion and frustration about the application procedures and specific requirements for LOAs, EFPs, and EAs. There appear to be differences between NMFS regions in terms of what constitutes scientific research covered by an LOA and what is covered by an EFP. In some cases there can be substantial delays in NMFS approval of applications, which result in project delays and, in some cases, their cancellation. NMFS must improve the clarity of the application procedure and the timeliness of the process. Similarly, scientific research leaders must follow the application process explicitly and submit applications in a timely fashion.

Charter Agreements or Contracts

The purpose of the charter agreement or contract is to explicitly state all the conditions and specifications related to the charter of the fishing vessel used in cooperative research. The request for proposals or quotations and the subsequent agreement should include similar elements. These encompass the period of the charter, costs and contingencies, vessel and operational requirements, and safety considerations.

The agreement must begin with a clear statement of the parties involved, the owner of the vessel, and the charterer. Specifically the agreement should address the following:

- The charter term or the period of charter and the geographic limit of the charter: Provision for foul weather should be considered in the agreement, as this may extend the charter period. The geographic limit of the charter should be within the limits covered in the insurance policy of the vessel and the safe operating range of the vessel.
- The requirements of the vessel: including length, tonnage, horsepower of the main propulsion engine; auxiliary systems (including electrical, hydraulic, potable water, seawater and waste, electronic navigational); fish-finding equipment; fishing gear (including winches, wire, nets, dredges, etc.); and deck and interior space (including the number of berths, showers, storage, etc.)
- Operational requirements: including the length of the work day and watch schedule, the number of meals, required assistance of the vessel crew to the scientific party, the disposition of the catch, etc.
- Crew requirements or the manning of the fishing vessel, specifically the qualifications and experience of the captain and crew
- Safety considerations, including the requirements that the vessel be equipped with all U.S. Coast Guard–required safety equipment, that all this equipment be inspected prior to commencement of the charter, that the vessel provide a stability letter demonstrating the vessel's stability report (and include copies of recent inspections and surveys that document the seaworthy condition of the vessel), and that the vessel provide appropriate safety equipment for all crew and scientific party members working on deck
- Provision for each party to terminate the agreement for justifiable reasons, but also specifying a penalty for early termination
- Fees, including the daily charter rate and the costs of mobilization and demobilization, foul weather or failed equipment standbys, and breakdowns
- The costs of fuel beyond some specified level, dockage, and modifications to the vessel
- Resolution of disputes using arbitration
- Integration, that is that the parties agree that the contract supersedes all prior agreements and discussions

Insurance

The type of insurance required depends on the nature of the cooperative research. In the case of sea samplers or observers aboard fishing vessels, the issue is problematic because it is not clear which insurance is applicable to this situation. State workman's compensation insurance is inadequate, as it does not apply to observers on vessels in federal or international waters. Longshore and harbor workers insurance, intended for shoreside workers, will not necessarily cover observers at sea. Jones Act insurance may not cover observers on fishing vessels, as they do not meet the requirement of "seamen" under the act. Companies hiring observers and placing them aboard fishing vessels often overlap insurance coverage, so as to be certain that they are insured. This increases the cost of observer programs. NMFS should provide clarification as to the source and level of insurance coverage to ensure coverage and reduce costs.

In the case of the chartering of a fishing vessel for fisheries research, the insurance issue is clearer. The operator of the fishing vessel should be required to procure and maintain indemnity insurance at a specified level, with the NMFS as a named insured, and the certificate of insurance should be presented prior to the commencement of the charter. The agreement should specify that the fishing vessel for itself and its insurers waive any right of subrogation against the NMFS, its employees, agents, or contractors for any claim arising for the vessel charter. Similarly, the NMFS agrees to defend and indemnify the fishing vessel for any claim or demand against the NMFS stemming from the operation of the fishing vessel, including the cost of defense. Any additional cost of the insurance should be included in the vessel charter fee.

If insurance will be necessary for a cooperative research project, adequate time and resources need to be allocated in advance of the project start date to prevent delays.

Data Collection and Enforcement

The success of vessel observers or sea samplers for use in collection of fishery-dependent data has been clearly demonstrated throughout many of the U.S. fisheries. An important distinction exists among observer programs: those that are mandatory and those that are collaboratively incorporated into the fishery (and are considered voluntary). Through cooperative observer efforts, the fishing industry and the scientific community can achieve a degree of ownership for data collected; trust in the validity of the information and invaluable collaboration can be established through working partnerships. These initiatives are not immune to problems.

Reporting of fishery violations by observers aboard cooperative vessels can present a dilemma. Mandatory observer programs have been utilized to jointly provide data collection and enforcement aspects. Voluntary observer programs have been designed more to collect fishery information and data. It appears that most cooperative observer programs are seldom exposed to violations; however the potential for encountering such problems exists.

Many cooperative observer programs exist in fisheries that incorporate relatively small or mid-sized vessels. In some cases, an observer accompanies a crew of three to four fishermen aboard vessels up to 85 feet in length for periods normally lasting from 30 to 45 days. The observer is placed in very cramped quarters with the crew and must live in close contact with them for long periods of time. It is stressed to the observer that teamwork should be established aboard the vessel. For cooperative research to succeed, it is vital that the professional fisheries observers live in harmony with the crew. The observer should be perceived as an unbiased collector of scientific information data and not that of an enforcer. Many potential problems can be avoided by clear agreement at the genesis of cooperative planning that fishery violations are not to be expected or tolerated.

It is a responsibility of the fishery observer involved in cooperative research efforts to record data as required, without regard to whether it could later result in a fishery violation. The observer generally has no obligation to report observed fishery infractions. However, if a vessel is cited for a fishery violation with an observer aboard, it is his or her duty to provide accurate information to authorities investigating the case. Furthermore, if the observer is summoned to court to testify as a witness for the prosecution, it is the obligation of the observer to report his or her observations truthfully.

Cooperative fisheries observers have a challenging occupation. They are often placed in a delicate and potentially contentious environment aboard vessels. Data collection can be a recondite task requiring special individuals who can adapt to rigorous conditions, yet maintain scientific integrity through use of prescribed protocols. If observers are perceived as enforcement adjuncts aboard vessels, a different atmosphere is established between the crew and the observer. This can result in the observer's primary job, data collection, becoming a much more difficult process.

Confidentiality of Data and Results

The data collected as a result of cooperative research projects or programs often contain information that is sensitive or confidential. Observers or sea samplers collect data ranging from catch rates and fishing locations to interactions with protected species. Release of this confidential information to other fishermen or to the press with specific reference to the originating fishing vessel can be problematic for the vessel owner, captain, or crew. All government parties involved with observer programs must make every effort to keep the data confidential, or when released they should be aggregated with those of other similar vessels so as to prevent the identification of individual vessels.

Often questions of immediate importance to fishery management or to the fishing community are being addressed by cooperative research. There is often pressure to release information or to speculate in a public forum on the implications of data as they are being generated. Data, observations, and results should not be released to the public until the project is completed, including the peer review process. This should be made very clear to all participants in the project. If the release of data or preliminary results will be necessary, this needs to be understood and agreed to in advance by all participants in the project. Such concerns are best addressed in contracts signed by all participants prior to the commencement of a project.

ENSURING SCIENTIFIC RIGOR

Standards Applicable to Cooperative Research

Cooperative research must meet the same levels of scientific rigor and quality expected of any other scientific research endeavor. Obtaining high-quality results depends on setting high standards and employing oversight processes to ensure that these standards are met. What makes cooperative research especially challenging in this regard is that if it is conducted in a highly cooperative mode, nonscientists who are generally not familiar with the scientific method are integrally involved in making decisions about the scientific aspects of the research project. Meaningful participation and buy-in of nonscientists in scientific decisions clearly require that some training and other mechanisms to facilitate understanding of scientific standards should be built into the project design and implementation process. It is also necessary for scientists to communicate with nonscientists so they can understand how they perceive marine ecosystems and the scientific process. Throughout the committee's deliberations it was emphasized that, in addition to passing scientific muster, cooperative research should be conducted in a practical, cost-effective manner and one that maximizes true cooperation and participation and the chances that the results will be used for management purposes. Thus, one often sees recommendations that cooperative studies be conducted aboard commercial fishing vessels operating under commercial fishing conditions (refer to examples in Chapters 2 and 3). Clearly, fishermen are the experts on the practicalities of employing various data collection methodologies (especially fishing gear) aboard commercial fishing vessels, and thus the scientific members of the team will need openness to the practical aspects of the project design and implementation. Scientists, on the other hand, are experts at utilizing and deploying an array of instruments according to rigorous protocols. Fishermen will need training that ensures that all participants understand the importance of these protocols and that there may be a need to compromise practicality and efficiency of fishing operations on occasion to ensure the validity and usefulness of the data being collected.

Ensuring that Standards Are Met

A cooperative research project team must include, at a minimum, fishermen and scientists. The committee heard reports of successful cooperative research in instances where the team scientists were independent of the management agency (e.g., academics), as well as instances where the team scientist was an employee of NMFS, although the former appeared to be more common than the latter. In instances where the scientist was an NMFS employee, the scientist did not have regulatory responsibilities. The process for developing, implementing, and evaluating the project should be broadly inclusive and transparent. Many of the cooperative research projects reviewed in this report made use of advisory committees with broader membership, including agency personnel with regulatory responsibility, science administrators, employees of environmental organizations, fishing industry association representatives, and additional scientists and fishermen. Such committees can provide an informal and consistent source of guidance throughout the duration of the research project. Such a committee also provides a ready mechanism for regular communication among project participants and managers, while maintaining independence of the research project from management pressures.

Cooperative research projects should be subjected to independent scientific and practical peer review at the proposal stage and at project completion. At the proposal stage, peer review should evaluate the validity and efficiency of the experimental design, data collection methods (including sample sizes, stratification, and data quality control mechanisms), and proposed data analysis methods. At the completion of the project, the findings and conclusions need to be peer reviewed. Where possible the findings and conclusions should be submitted to a recognized scientific journal for publication. For multiyear projects, one or more interim reviews are advisable to be sure the project is on track and meeting its intended objectives. Programs that administer cooperative research should also be subjected to programmatic peer review not less often than every five years.

Validation of data accuracy and precision is an important part of the research process and overall project quality control. Every cooperative project needs a system of data verification and quality assurance. Independent scientific observers aboard commercial fishing vessels are likely to be the major tool employed for data quality assurance, but other surveillance tools may be useful in some circumstances.

Procedures regarding the announcement, use, and publication of results should be clarified as part of the initial contractual agreement between the project team and the cooperative research-granting entity. In all cases, following project completion, results should be disseminated broadly and be available in the public domain.

6

Incentives and Constraints to Cooperative Research

The primary institutions and participants from which the committee heard information were the National Marine Fisheries Service (NMFS), fishing industry organizations, and academic participants in cooperative research. Other institutions that either participate directly or have a say in funding, plans, and results of cooperative research include the Congress, the regional fishery management councils, Sea Grant, the states, private science institutions, recreational fishing organizations and individual anglers, and environmental groups and their members.

Each of these sectors has its own decision-making processes, planning schedules, governance structures, financial arrangements, and motivating interests. All of these factors can influence how cooperative research projects are conceived and carried out. In some cases, administrative structural issues can become constraints to cooperation even when motivation to collaborate is high.

Cooperative research entails the interaction of fishermen and scientists at both personal and institutional levels. A common theme in presentations to this committee was the importance of institutional attributes and individual personalities in the success or failure of cooperative research projects.

On an institutional level, commercial fishermen participate in cooperative research through industry organizations and private businesses, as crew members and boat owners. To a lesser extent, recreational fishermen, conservation advocates, or environmental groups also participate. Scientists engage in this work through their positions at universities, state natural resource agencies, federal agencies (such as the NMFS), extension services, and as industry or environmental group consultants.

Institutional cultures and structures can inhibit or facilitate cooperative research. An organizational culture that is flexible, adaptive, and open will be more likely to promote cooperative research than one that is rigid and closed. Structural variables include funding, staffing levels and expertise, organizational mandate, personnel policies, and bureaucratic requirements. Organizations must be structured so that the extra time, resources, and flexibility required to conduct cooperative research can be accommodated.

Incentives play an important role in cooperative research. Positive motivations for scientists and managers include professional advancement, additional resources for getting necessary work done, improved information for decision making, and better relationships among fishermen and scientists. For the industry, better information, increased understanding of science, potential for additional catch, and cash or in-kind (additional quota) payments provide incentives. Negative motivations can also be effective in the support of cooperative research. Fear of closures, quota reductions, penalties for bycatch, and loss of income can be strong motivators for fishermen to participate in cooperative research projects as well. This chapter will examine the incentives and constraints in NMFS, academia, and the fishing industry that affect the success of cooperative research.

NATIONAL MARINE FISHERIES SERVICE

Structure and Culture

NMFS is a decentralized and complex agency. It has resource management tasks from enforcement and seafood inspection to ecological assessments and protection of endangered species, from gear technology development and fishery management to resource surveys and stock assessment. These jobs are carried out in eight distinct offices, within headquarters, five regional offices, and five science centers. Eight regional fishery management councils and three interstate fisheries commissions have interwoven roles in both management decision making and the planning and implementation of research.

NMFS has a number of difficult and potentially conflicting roles including regulation, science, conservation, and management. How do these roles affect the agency's ability to conduct successful cooperative research?

Is fisheries collaboration a discrete program or a philosophy of conducting improved fishery science? If it is a discrete program, are projects ad hoc or will they be conducted according to long-term systematic procedures? On the other hand, if cooperation is a philosophy, what must the agency do to develop the kind of values in NMFS scientists and administrators that foster collaboration? What incentives can it provide to encourage and support cooperative research?

The committee heard from many participants that NMFS's regulatory and enforcement responsibilities infiltrate and undermine cooperative data-gathering efforts in several ways. Where industry mistrust of NMFS runs deep, cooperative research may be perceived as a maneuver to promote a predetermined policy. Despite the fact that industry is directly involved in the design and implementation of many cooperative research efforts, a consistent thread running through presentations to the committee was the concern that the data from such efforts could be used "against" them because industry has had little voice in the interpretation of the data.

The role of the regional fishery management councils (FMCs) in requesting, designing, and considering the results of cooperative research needs to be clarified. In some regions, FMCs were partners from beginning to end. In others their role was seen as too political and not objective. Yet the committee heard from most presenters that the information needs of decision makers should guide development of research priorities and funding. However, it also was pointed out that the constraints and requirements of the regulatory system (when annual quotas are set, FMC schedules, statutory decision deadlines) can complicate, delay, and defeat cooperative efforts.

Another concern was that pressure on NMFS to develop outreach activities and improve its constituent relations could either foster good cooperative research or undermine authentic and scientifically rigorous collaboration with superficial public relations projects (the so-called Discovery Channel projects). Another danger that was pointed out lies in the kinds of projects that are mandated by specific budget line items and are widely viewed as cash assistance to fishing communities rather than scientifically valuable research projects.

Finally, there was concern that the culture of a science and management agency like NMFS does not necessarily seek out, hire, train, foster, or reward the skill sets or attributes that contribute to successful cooperative research.

Administrative and Fiscal Constraints

A primary constraint to NMFS participation in cooperative research is its inability to control the funding cycle for cooperative research. The time lines for funding, management needs, data requirements, and priority setting are not aligned, making it difficult to plan accurately. The administration of cooperative agreements is a long process. It can take five months to contract with a fishing vessel, and field offices need to start a year in advance to get administrative paperwork completed before beginning the research.

Not only do field offices not control their funding time lines, they do not know when to expect funds to be dispersed. The committee was told of instances in which funds were received in the middle of a project year. Such lack of control makes scientific staff hesitant to plan with industry because they don't know when funds will arrive. This can erode trust between NMFS and fishermen if an agreement for work has already been made. The expectations of the fishery are raised and then NMFS is unable to carry through.

Potential conflicts between fishermen and agencies in arranging survey schedules to accommodate fishing season openings have been resolved in regions with successful cooperative relationships. In other areas they have doomed projects.

Even more significantly, the agency may not even have much say in the content of cooperative research. Often line items for specific projects appear in the NMFS budget because constituents in the fishing industry have told their members of Congress what they want to see funded. On the other hand, industry has worked with Congress to expand NMFS's base budget for research and stock assessment as well as for cooperative research.

Another fiscal constraint is that only 15 percent of the funds appropriated for cooperative research can be used within the agency by NMFS scientists. Cooperative projects require substantial staff time commitment and infrastructure to support them. The committee heard that there are not enough scientists to handle the current research workload, so cooperative projects come as an additional burden, even at the cost of basic activities. Managers feel that cooperative projects have to be relevant to already-stated research needs and can be undertaken only where they have staff to participate.

Incentives and Disincentives to Cooperation

The main incentives for agency participation in cooperative research appear to be improved relationships with industry, increased credibility and acceptance of biological information, and the addition of research platforms, practical expertise, and manpower. Mandates, of course, provide clear motivation to do cooperative research but also cause resentment.

An incentive that does not exist at present but was suggested to the committee as a positive way to encourage collaboration is to reward NMFS employees and scientists for participation in cooperative research. This could be done by:

- recognizing the work as a path to professional advancement;
- counting the publication of the results of such projects as part of professional development; and
- providing opportunities for training and tradeoffs relative to traditional research, rather than making cooperative research projects an additional burden.

In contrast, the disincentives to NMFS scientists to participating in collaboration with industry partners include:

- cooperative research may be used as a vehicle by which to prove NMFS wrong;
- improperly planned or unsuccessful cooperative research projects may divert funds from necessary projects;
- emphasis on collaboration may limit dollars for most effective research (and may limit research on long-term or politically unpopular issues such as habitat or bycatch);
- a shortage of scientists for collaboration; and
- a lack of administrative and infrastructure support for cooperative research projects.

There are far more fishermen than scientists willing and able to enter into cooperative projects because the incentives are present for them (see below). This imbalance, coupled with the increasing insistence of government to enter into such partnerships without increases in government science money, results in high workloads and stress levels for scientists. Many are running a number of cooperative research projects simultaneously.

ACADEMIA

In general, neither the socialization into science that students receive while in college or the training that scientists receive from the institutions they work for promotes cooperative research.

First, as a rule science is competitive and not cooperative. Students compete to get into graduate school. They compete for fellowships and grants and for their major professor's time and attention. Junior faculty compete for grants, publication space, and tenure. They not only need to publish, they should be first author on most of their publications and principal investigator on most of their grants. Both graduate students and faculty are rewarded for original ideas and the expansion of theory. These are individual pursuits, and recognition in the form of tenure and promotion is for individual contributions.

Second, most research universities promote basic over applied research. In presentations to this committee, some fishermen worried that scientists are rewarded for "hotshot" science, not "roll up your sleeves" science. There was broad agreement that there are not enough field-oriented scientists to do the kind of research fishermen would like done.

Third, faculty face time constraints. Research is only one of three academic duties, the others being teaching and professional service. Teaching loads can vary from one to four courses a semester, depending on the university and one's position. Cooperative research can be more time consuming than "traditional" research because a good working relationship with fishermen is key to its success. This requires numerous meetings, visits to geographically dispersed ports, and phone calls. The pressure to publish in peer-reviewed journals and present at professional meetings can place additional time burdens on cooperative researchers who are also expected to write peer-reviewed project reports and present results to management and industry partners. It can force them to choose between publishing venues that are highly regarded by university administrators but have no management impact and venues that are not highly regarded by university administrators but will have a management impact.

Fourth, scientific training does not adequately prepare students to do cooperative research. As mentioned above, students are socialized to compete rather than cooperate. In addition, they are taught that an understanding of the natural world must be based on information derived from the scientific method. The scientific method is a very particular way of generating knowledge based on observation, hypothesis testing, and the ability for results to be replicated. It does not easily incorporate the experiential and anecdotal knowledge of fishermen.

Changes in the reward structure in academia need to take place if more academics are going to participate in cooperative research. The current prestige structures that place higher value on peer-reviewed journals than on peer-reviewed reports and white papers need to be reevaluated and contributions to management and policy need to "count" in tenure and promotions equations.

INDUSTRY

Current depressed conditions in a number of fisheries around the United States and the accompanying management measures that limit catches, seasons, and effort have provided a strong motivating factor for fishermen to participate in cooperative research, whether as a supplemental income source to fishing or as work that has a direct bearing on quotas, allocations, and seasons. Throughout discussions it appeared that industry members who were very much engaged in cooperative research felt they had a stake in the fishery and believed that by participating in research they were contributing to research that may help conserve fishery resources and allow them to fish with reduced impacts on nontarget and protected species.

Structure and Culture

By definition, cooperative research involves participants from groups with different priorities, histories, traditions, backgrounds, and standards of behavior. In some instances, these cultural differences contribute to miscommunication and misunderstanding, mistrust, conflict, delay, and the real possibility of failure. Case studies and comments to the committee clearly show that the differences in cultures among professional researchers and scientists and fishermen must be recognized and dealt with if cooperative efforts are to succeed.

Although fishermen do form associations, marketing cooperatives, and other organizational structures for collaboration, the business, at its core, is a solitary pursuit. Independence and a lack of cohesion of industry sectors are of long-standing tradition. The committee heard in all regions about the difficulty in communicating with and engaging a majority of any fishing sector or gear group and the importance of leadership in accomplishing them. Confidentiality is an important aspect of competitive advantage, and collaboration may be viewed as a threat to that privacy.

It also needs to be acknowledged at the outset that fishermen are the "regulated community" and NMFS is the "regulator." The commitment required to overcome the adversarial aspects of this relationship and to move into a collaborative relationship is not insignificant, but there are ways to do it. Time at sea is highly valued and can be a basis for building trust. But time spent at sea fishing is different from time spent conducting research. Fishing culture, where time is money, calls for a much faster rhythm and pace than do scientific surveys. However, presenters appearing before the committee and reports in all case studies emphasized the trust building that resulted from time spent in a project at sea solving problems.

Not enough can be said about the importance of long-standing relationships of the type that historically were built when NMFS personnel spent more time on the docks talking to fishermen personally. The committee heard from both fishermen and agency officials that cooperative projects were more likely to be undertaken and to succeed where agency personnel had not just an open door policy but a policy of truly listening to fishermen.

Finally, fishermen evaluate and interpret evidence very differently from scientists. Efforts to produce statistically significant results look like redundancy to fishermen, who are more willing to depend on their personal observations and experience and to base conclusions on smaller amounts of data and different types of information.

Administrative and Fiscal Constraints

Fisheries management is an unavoidably bureaucratic system. Most of the cooperative data-gathering efforts the committee heard about became enmeshed in bureaucracy at some point in their history. Whether legal concerns about confidentiality, rules of the Paperwork Reduction Act, insurance, permits, or timing, all these examples pointed out the importance of a serious commitment, staying power, and ability to navigate the system on the part of industry leaders. Financial issues are not simply administrative or legal, however. Whether it is a matter of contracting with a vessel as a research platform or working with fishermen from the inception of a project in its design to its implementation, money is tied up in vessels, equipment, crew, and the skipper's own time.

One industry representative told the committee that prosperous fishermen make better partners for NMFS than desperate ones do, and in many cases, participants in cooperative research brought significant capital to a project that enabled it to go forward in ways that could not have happened with agency resources alone. However, even prosperous fishermen need to have some business certainty to participate in a cooperative project.

One concern was that NMFS does not disburse money quickly or on schedule. This can pose a problem for collaborating partners because fishing businesses are very sensitive to cash flow, don't carry large lines of credit, and will avoid relationships where payment is delayed. Some fishermen the committee heard from encountered problems with the grants process. They noted there is often competition for funds between the scientific community and industry, making it difficult for fishermen to put in a proposal that is competitive. The committee heard information on numerous grants options, each one with its own rules, deadlines, requirements for review, and so on. It would help the process of finding collaborating partners if there were a clearinghouse, fact sheet, or common place where prospective partners could examine various options for receiving grants, contracts, experimental fishing permits, total allowable catch research set-asides, and similar awards. Assistance in putting together proposals and assurance that the grants process is open for everyone and that there is a level playing field where industry partners can compete are also important elements.

Incentives and Disincentives to Cooperation

Fishermen need incentives to participate in cooperative research, and they have to be significant to overcome the inertia of not participating. Incentives can include compensation, more fishing time, catch to sell, changes in the information decision makers use that could affect management measures, or a better understanding of scientists and the scientific process. In addition, the committee heard of instances where cooperative projects provided recognition and empowerment to fishermen.

For example, questions that fishermen have had with assessments or management decision issues can be addressed and then proven or disproved. One commenter told the committee that fishermen are seen as more valuable to society when they are data gatherers as well as producers of seafood. Fishermen who participate in cooperative research projects get scientific products that have more credibility because they contributed to them. Fishermen want to use their knowledge of fishing in the management sphere, and cooperative research is a mechanism for legitimizing that knowledge.

Another incentive for participating in cooperative research is the fact that there is so much information that must be collected that additional participants and vessels are necessary to fulfill the need. This may become a problem in areas like the West Coast, where fishery closures and catch restrictions are driving the downsizing of the trawl fleet and the kind of leaders that make collaboration work are taking other jobs and no longer fishing. An agency official told the committee that as fleet size declines it is more and more difficult for captains to find quality crew members (who usually exit first).

Yet another vigorous incentive is the requirement for bycatch reduction. Fishermen in several regions have tested gear, applied for experimental fishing permits to develop gear, and collaborated with NMFS gear specialists to test gear the agency develops. Not only do these efforts contribute to knowledge about bycatch and fish behavior, the development of gear that reduces bycatch allows some fisheries to stay open longer. These projects also provide a venue for give and take between the agency and industry and an outlet for the kind of practical advice that fishermen have developed from years on the water.

Disincentives and Constraints

Although it is less true today than it was as recently as the mid-1990s, cooperative data gathering is still seen as a change from the status quo, in which NMFS has the principal responsibility for data collection, analysis, and interpretation. Overcoming the financial risks, practical impediments, and bureaucratic obstacles of cooperating with the regulatory agency requires strong motivation for fishermen. The disincentives are many.

The committee heard that cooperative research is sometimes looked upon as nothing more than disaster aid, putting fishermen in the position of being seen as a drain on public resources. Even if a project is significant, working on cooperative research can sometimes mean making less than if the same vessel and crew were at work fishing. And just as for agency scientists, fishermen face the risk that a data collection project could prove them wrong.

OTHER CONSTITUENT GROUPS

Although cooperative research is most often conceptualized as involving agency and university scientists with commercial fishermen, other stakeholders, such as environmental groups, recreational fishermen, and even former NMFS and academic scientists have also participated.

Environmental Organizations

Several presenters to the committee stated that environmental organizations should be involved as partners in cooperative research. To date, their involvement has been limited. Bernstein and Iudicello (2000) analyzed six cooperative research projects. Of these, two involved environmental groups. Out of eight case studies submitted to the committee for inclusion in this report, none involved environmental organizations. In those cases involving environmental organizations, environmentalists were involved primarily via membership in advisory stakeholder groups. Another way in which environmental groups have participated in cooperative research is as intermittent observers of the process on research voyages. This happened in a limited fashion in the Gulf of Mexico and Pacific. An exception to this limited participation occurred in Hawaii, where the National Audubon Society helped conduct research and write a research report.

The limited participation of environmental nongovernmental organizations in the actual execution of the research cannot be explained by their lack of scientific expertise. Many employ scientists with advanced degrees. It is important to examine potential causes of their lack of participation given the importance of participating in fieldwork for the construction of trust among partners (Bernstein and Iudicello, 2000).

Structure and Culture

There are two types of tension inherent in many environmental organizations that can erode trust between scientists and industry on the one hand and environmental organizations on the other:

1. The potential tension between achievement of conservation goals and the promotion of the fishery

The environmental community is diverse. Organizations differ in their approach to the use of natural resources, their use of scientific information, and their preferred mode of action (e.g., litigation, research, lobbying, stakeholder groups). Despite this diversity, environmental organizations have at least one thing in common. They were formed to promote conservation. Although many environmental groups want to preserve commercial fisheries, they are generally more willing to support short-term limits to the fishery in order to make long-term gains. In the bycatch excluder case study presented in Bernstein and Iudicello (2000), several environmental organizations that were members of the cooperative group were suing NMFS for failing to curb overfishing. In another case, fishermen felt that the environmentalists used the media in a manner that harmed the collaboration. Actions like these contribute to a climate of distrust between fishermen and environmental organizations. Fishermen feel that pressure from environmental groups contributes to stricter regulations.

Environmental organizations, however, can aid in the promotion of cooperative research through their endorsements. Environmental organizations supported testing industry gear as well as NMFS-designed gear in turtle excluder device (TED) experiments, thus opening up the range of TEDs to be examined.

2. The tension between the promotion of sound science and activism

Environmental organizations provide public education and a venue for public activism. Fishermen have complained about the misrepresentation of fishery issues and fishermen by environmental organizations. This can lead to questions about the neutrality of scientists from environmental organizations and their ability to participate in a cooperative research effort. Like NMFS, environmental organizations' dual mandate can erode trust.

Administrative and Fiscal Constraints

An environmental group might not be able to participate in cooperative research because of its own limitations. Lack of scientific staff with the appropriate background limits an organization's capability to conduct or participate in research. There are only a few large organizations with a diverse fisheries scientific staff, although some smaller organizations do have staff dedicated to research and are currently participating in cooperative research. Most environmental organizations identify a few key initiatives and hire staff with expertise in these key areas. If a cooperative research project does not fall into one of an organization's campaigns, it might be unable to fully participate. This could be one reason that environmental organizations seem to participate through committees rather than in the execution of projects.

Incentives and Disincentives

The incentives for environmental organizations to participate in cooperative research are related to promoting sound conservation of fishery resources and the marine environment as well. Through participation in cooperative research, they can promote their goals while building trust with industry and NMFS.

Disincentives might come from members. Most environmental groups rely on members for some degree of funding and volunteer activities. Members hold a range of views about fisheries issues and some may view collaboration with fishermen as co-optation. There is a lot of pressure in multi-stakeholder collaboration to reach consensus, and organizations losing sight of their original goals during the course of collaboration have been documented (Wondolleck and Yaffee, 2000). Therefore, environmental groups can risk their credibility when they participate in cooperative research.

7

Outreach and Communication

Outreach and communication are vital elements in the success of cooperative research programs. It is important that the fishing industry have a degree of ownership and commitment for research activities in which it is involved. The industry should have a clear understanding of the research projects. Furthermore, results of cooperative research should be clearly communicated to the resource users. There should be a continuous feedback loop of information to all participants in cooperative research. Often the weakness in a cooperative research project stems from differences in professional experience and expectations of the partners. To succeed, each project must develop its own internal "working culture" and pay attention to that from the outset. Both sides have to do "homework" regarding the other's turf to obtain a clear understanding of the partnership. Although the National Marine Fisheries Service (NMFS) has traditionally maintained excellence in science, efforts regarding outreach and dissemination of research results have often been inadequate and ineffective. Fishermen who pursue cooperative research tend to be very professional and may have as much or more of a professional stake than their science partners. This problem becomes more recondite with the diverse fisheries, cultures, and regions within the responsibility of NMFS.

COMMUNICATION

Many scientists have no problem communicating with fishermen, while some researchers lack the innate ability to do so. Because a broad spectrum of communication is such a necessary part of cooperative re-search, emphasis should be directed toward effective expression. Industry must have a clear understanding of project goals, scientific protocols, and research results. Both scientists and fishermen must be able to clearly com-municate problems and concerns that might often be encountered with research activities.

To assuage possible communication problems, NMFS either needs to invest in developing these capacities in-house, or it needs to build effective relationships with entities that do. State agencies, academic institutions, Sea Grant, and nongovernmental organizations (NGOs) are obvious part-ners for improving communication. Another approach might be to employ fishermen who have the respect of all parties and the skills to communicate both with the fishing industry and the scientific community.

Lack of understanding of the social structures of various fishing com-munities by NMFS was identified as a problem. Cooperative research is an ideal vehicle for NMFS to extend a network of positive relationships into the fishing community. It is a means to identify and establish channels of communication with local "community leaders." Emphasis should be placed on improving NMFS's knowledge of the fishing communities. Internally this can be done through interaction with fishery reporting specialists (port agents), who are located in many strategic ports or through the current expansion of social scientists now being employed by NMFS. Externally, key industry leaders, Sea Grant, NGOs, and state agencies can be valuable resources in defining the structure and nuances within various fishing communities.

Understanding and communicating with various ethnic cultures within the fishing communities often require different techniques and approaches. In some regions, churches have provided access to certain community sub-cultures (Vietnamese and Sicilian), whereas in other cases industry organi-zations, such as the Vietnamese American Shrimp Association, have pro-vided important assistance and knowledge.

The most effective, though least efficient, method for communicating with fishermen is one-on-one contact. Because fishermen must often re-main at sea for lengthy periods of time and then perform chores on their vessels while in port, many fishermen do not attend organized meetings. For these fishermen, one-on-one contacts with other fishermen on the waterfront provide the best forum for communication; however, it is time consuming and costly. The extent of individual contacts often is limited by budget and staffing issues. Despite these obstacles, one-on-one conferences should be encouraged when possible.

OVERCOMING THE PERCEPTION OF ARROGANCE

The institutions of commercial fishermen and the institutions in which academic or government scientists work could not function more differently. The cultural milieu of the two disciplines is antithetical in many ways. The rhythms of fishermen's days are tied to weather or tide and place them out of sync with the workaday world of most Americans. And though fishermen and scientists may speak the same language, they don't use the same vocabulary. When a fisherman says that "scientists are arrogant" or when a scientist feels that fishermen are "aloof and uninterested," it may be that cultural differences are getting in the way.

Bringing fishermen into successful partnerships with scientists often means that there has to be some flexibility in scheduling and in choosing times to communicate. Government or academic institutions embarking on cooperative research projects have to employ people who are sensitive to these irregularities, are adaptable, and are prepared to work in the evenings or whenever a fishing partner can be available.

Commercial fishermen want to be respected for what they know and for the information they provide to be recognized as being valuable. Scientists and outreach people who are the most successful at working with the fishing community treat fishermen as if they had earned a college degree in life experience. In these situations, some attention to interpersonal relations goes a long way toward breaking down barriers to true communication and learning.

Both scientists and fishermen should constantly be kept aware of differences in the way they use words and the meaning of words. Fishermen should be encouraged to explain their gear, techniques, or observations in detail. They should be challenged to communicate clearly and held to high standards of precision and accuracy in reporting. Scientists need to be wary of jargon and need to think about how to express concepts in plain English while avoiding the appearance of condescension.

OUTREACH

Outreach greatly determines the success and perception of cooperative research. Although cooperative research can serve the purpose of science, it is also a natural fit for NMFS to develop and expand its outreach activities. Disseminating results from cooperative research projects provides an excellent opportunity for establishing communication within the fishing industry. Historically, NMFS has not disseminated information well. Successful outreach efforts are contingent upon disseminating information to fishing communities in a manner that reaches user groups in a timely fashion and is understood by the layperson.

A well-designed cooperative research project should make a clear distinction between data gathering and data analysis and provide for outreach at both phases. Outreach during the data-gathering phase prepares the community for the work, often simplifies the logistics, and puts the "hypothesis" in plain terms. Often, outreach during the course of the research, especially with gear-related development, will precipitate ideas that focus or improve the research. Outreach at the analytical phase often employs fishermen in "making some sense" of the information, attempts to dampen speculation on the data by putting them in the context of the experimental design, and informs the community. Care must be taken, though, not to reveal the results or conclusions until completion of the research project, which includes the peer review process. For longer projects, the reporting of interim results (following peer review) may be appropriate. This should be agreed to by all participants before initiation of the project.

While the results of cooperative research should be released to all at the same time, dissemination of results can be enhanced through use of select fishermen who have the knowledge of scientific principles, communication skills, industry respect, and motivation to perform educational activities. These individuals can be employed to perform one-on-one contacts with industry within the fishing communities as well as through traditional educational forums, such as workshops, seminars, and so forth. In addition to the utilization of specialized fishermen to perform informal outreach efforts, industry collaborators who have participated in cooperative research activities can be (and have been) effective educators within the community. In addition to regular contacts with their peers, industry investigators can be utilized in educational forums to communicate project results. Not only does this outreach present an opportunity to enhance communication within the industry, it serves to instill ownership of the cooperative research project and results within industry.

A number of effective educational methods exist for outreach. Historically, many of these have been applied with excellent success through the land grant institutional process. A list of methods would include:

- One-on-one contacts
- Community workshops and seminars
- Presentations at regional fishery management meetings such as fishery management council meetings
  - Presentations at NGO meetings and conventions
  - Presentations at trade conventions
  - Presentations at fishermen forums
  - Regional and area outreach forums
  - Newsletters
  - Extension publications
  - Web sites
  - Educational videos
  - Local newspaper articles
  - Trade periodicals
  - Formal reports
  - White papers

While Web sites can serve as an educational tool for cooperative research projects, they need to be constructed in an interesting and user-friendly format. The use of video footage has proven to be an excellent educational tool. In addition to utilizing videos during presentations of data, industry has demonstrated that educational videos will be individually utilized, both at home and at sea, when distributed to fishermen.

When possible, visual aids should be incorporated into outreach presentations. In some regions, illiteracy still exists among some fishermen. In certain areas, a large number of fishermen are not fluent in English. The use of interpreters is often an asset in disseminating information in these areas.

There are numerous templates for outreach activities that have been conducted throughout the United States over the past several decades. Many of these have been developed and employed by universities through Sea Grant and cooperative extension activities. Cooperative research provides an opportunity for NMFS to partner with these organizations and to expand its effectiveness in communication and outreach.

THE ROLE OF TRANSLATORS

Fishing cultures are verbal cultures. The "grapevine" is still the most trusted source of information. A truism in the fishing business is that "the three fastest forms of communication are telephone, telegraph, and tell-a-fisherman." It is not unusual for a piece of information to be relayed over thousands of square miles within hours through "chatter" on marine radios. Like the Internet, misinformation is just as likely to take on credibility as information.

Typically, individuals listen to the grapevine, or the marine radio, as background to their daily work. Multiple conversations on multiple channels are monitored simultaneously. One conversation may be dropping hints as to where the fishing is good, another speculating on market trends, and still another conversation discussing the latest management action. Some fishermen are more active broadcasters than others, some are very respected, and some do not speak at all. From this ongoing fleet-wide conversation, the collective community culture forms and perpetuates opinions. Cooperative research activities and information tend to resonate strongly within this broadcast field. Scientific research being conducted on the deck of a fishing boat is something fishermen will talk about on the marine radio. These conversations will not only be about the logistics of the research being conducted but the implications as well.

Because cooperative research tends to attract fishermen who are innovators, many of these people are already recognized leaders in their communities. In their new role as a researcher, many serve as translators of information. It is extremely important that cooperative research projects not underestimate the fishermen partners' ability to inform. The more that fishermen are integrated in the design and experimentation, the more familiar they will become with the scientific method and the analytical tools. This will in turn increase the likelihood that the information that flows to the community is accurate and complete.

One of the most important results of cooperative research is the emergence of translators. These are people from the fishing industry, NGOs, Sea Grant, and sometimes state or federal science agencies, that operate on the cultural interface. Fishermen in the role of translator serve a vital function in the increasingly complex world of fishery management. Firmly rooted in the community's values, they are trusted sources of information. They help interpret management action and can help direct their fellow fishermen through bureaucratic snarls.

8

What Works and Doesn't Work

To this point, this report has provided numerous examples and discussed a variety of issues concerning cooperative research. In Chapter 1, the topic of cooperative research was introduced. In Chapters 2 and 3, case studies were provided to illustrate some experiences with cooperative research both in the United States and in other countries. In Chapter 4 the setting of cooperative research priorities and processes, along with descriptions of potential mechanisms, was discussed. Funding, legal issues, and scientific rigor were discussed in Chapter 5. Chapter 6 dealt with the constraints and incentives and disincentives for cooperative research and Chapter 7 with issues related to outreach and communication of cooperative research. In this chapter the information from the previous chapters is summarized, with focus on what works and doesn't work in cooperative research.

REASONS FOR SUCCESS

Cooperative research works when scientists and fishermen realize that each bring valuable tools and experience to the objectives of a research project. Scientists who are successful in cooperative research realize that fishermen have knowledge, skills, and/or vessels that would not otherwise be available and are willing to work with fishermen in order to get the desired results. The fishermen in these successful projects are also willing to work with the scientists, recognizing that the information collected will not be used in decision making unless it is scientifically credible.

Fishermen involved in cooperative research often acquire a much greater understanding and breadth of vision than provided by their fishing experience. When a fisherman becomes a partner in cooperative research, he/she often learns to value the scientific method and is better able to differentiate valid science from speculation. For a fisherman, having an open mind often means putting his fishing knowledge temporarily "at arms length" while learning about the science objective and the analytical tools involved with the work and then synthesizing his own experiences with that new point of view.

Just as fishermen can achieve a new perspective on scientific research, scientists can also learn to value the knowledge of fishermen. When cooperative research engages fishermen as experts, the information flow will not just be from the scientists to fishermen, but scientists will learn the value of enhancing their knowledge with the perspectives of fishermen.

Participants in cooperative research projects have explained that the whole of the participants' knowledge is often greater than the sum of its parts. That is, ideas and understanding arise where fishermen and scientists work together in a single project that neither group would develop on its own through fishing or fishery-independent research.

A prerequisite of successful cooperative research is that all participants thoroughly understand that they are involved in scientific research, even though fishermen may be responsible for much of the design and execution of that research. It must meet scientific standards if it is to be useful for management decisions. Cooperative research projects must apply scientific rigor with the same standards that are applied to traditional (dedicated) research if they are to have credibility. From the inception to its conclusion, the supervisors of cooperative research projects need to emphasize this need to all participants and make sure the project will produce scientifically defensible results.

Cooperative research projects have worked when the participants fully cooperate from the start of the project to its finish. In the initial step, fishermen and scientists acknowledge that a problem or opportunity exists that needs to be addressed. Second, they determine that working together cooperatively is the most effective means to solving it. They see that the involvement of fishermen along with the scientists is essential to success of that research. Third, they use each other's expertise in science and on the fishing grounds to design the most appropriate and practical research protocol.

Fourth, they execute the project according to the original plan and within scientifically valid guidelines. They do this by extensive communication among all parties at each step of the project. Finally, they properly interpret and distribute the results of their project to the affected parties and management agencies. Likewise, the government, or the regulatory agencies of the fisheries involved, also agrees on the cooperative process as the best method to solve a recognized problem. The agency supports the project to its conclusion, both administratively and financially.

In Chapters 2 and 3, examples and case studies of cooperative research illustrating a broad range in levels of cooperation were provided. The examples and case studies ranged from no government participation (except in the review process) to industry taking the lead on developing cooperative research but where government scientists participated in the data collection, to cooperative research dominated by government scientists in planning and execution but including fishermen in some way in the execution of the project. There does not appear to be any "best" way. Different approaches have been successful for different types of problems in different institutional settings, but the overriding theme is the desire to provide improved information for decision makers that meets accepted standards of design, program execution, and analysis.

The Motivation of Fishermen

There are a number of reasons why fishermen may be motivated to participate in cooperative research. Fishermen may perceive a threat to their fishery from pending management action. When dolphin bycatch mortality in the tuna seine fleet precipitated widespread boycott of tuna products by the general public, eastern Pacific tuna seiners were motivated to develop dolphin avoidance techniques in conjunction with the Inter-American Tropical Tuna Commission, starting in the 1980s. Potential sanctions of the Endangered Species Act likewise threatened to close the North Pacific longline fishery in the late 1990s due to bycatch of an endangered species, the short-tailed albatross. Fishermen did not have to be convinced that this was a problem and to seek a solution.

Other projects arise from financial motivations. In 2001 a mid-Atlantic surf clam fisherman thought that the government used faulty stock assessment techniques. If the current level of effort was resulting in underharvest, there was additional resource available to the fishermen. If the current level of effort was resulting in overharvesting, the resource in which they have a long-term stake could be in jeopardy. In this case, the fishermen approached the National Marine Fisheries Service (NMFS) and convinced them of the problem and the need to work together to find a solution. The development of the New Zealand rock lobster logbook program was similarly motivated by a desire of the fishermen to "prove" the status of the resource. Individual fishermen's observations or logbooks were not considered adequate as input to the stock assessment process, so the fishermen arranged for a scientifically designed program that now forms the basis of the stock assessment.

Another motivation, one most easily misunderstood or easily dismissed, stems from the fishing community's desire to investigate alternate hypotheses (to what management or scientists may suggest) or simply a desire to improve the scientific information used in managing fishery resources. Sometimes the "best available information" may be significantly improved upon at an acceptable cost. That cost-benefit analysis has to be conducted by fishermen and scientists together. For the fishing community it is an issue of self-determination. In the northeast United States stock boundaries for codfish were established decades ago from tagging studies, growth rates, and parasite loading. Currently, management recognizes a Georges Bank cod stock and a Gulf of Maine cod stock with distinctly different biological reference points and management strategies. Many fishermen contend that mixing rates are a significant factor at different life stages and confound these established stock models. Different stock boundaries or a more complex stock model would alter the present management regimen. A series of port meetings documented fishermen's observations, and a task force was convened to design a comprehensive tagging program. Starting in 2003, this program will be carried out with cooperation of fishermen from four states and the Canadian maritimes.

The Motivation of Scientists

There are often far more fishermen than scientists willing to participate in cooperative research. Cooperative research is often seen in the academic community as not being "legitimate" research. In addition, scientists involved in cooperative research must make additional commitments of time and effort to foster good working relationships and true cooperation with their fishermen partners. Agency and university scientists need to publish if they want promotion and tenure. If cooperative research requires a greater time commitment and is difficult to publish, many academic scientists will not participate. Agencies and academic departments involved in fisheries research need to take account of these challenges. This can be done by not using the same evaluation process for promotion and tenure that is used for research faculty and staff not involved in cooperative research: one that accounts for the additional time commitments and impediments to publishing.

Another possible remedy would be the establishment of a scientific journal of cooperative research. This option could provide an incentive for more scientists to participate in cooperative research, but it could also serve to further marginalize cooperative research from traditional research. Such a journal would need to have a rapid turnaround time, and it might be possible to have the peer review of the cooperative research projects serve as the peer review for the reports and the reports then published online with rapid turnaround. An alternative is to encourage scientific journals to publish more cooperative research.

Scientists who have done successful cooperative research often state that the satisfaction comes from factors other than producing publications. The scientists may simply enjoy life on the ocean and time away from the office or lab. They may simply like to work with fishermen or seek the satisfaction of working on cooperative research and knowing that a real and immediate problem is being solved through the research. The results of a cooperative research project could provide the basis for changing laws or regulations, prevent a fishery closure, allow a fishery to function without onerous sanctions, or conserve fish and other marine life.

Scientists can have a financial motivation as well. Chartering a fishing boat may be less expensive than chartering a dedicated research vessel. They may choose to charter a fishing boat because the cost may be charged to different funding sources or can be funded by catch. For decades the International Pacific Halibut Commission has chartered fishing vessels with fish caught on charters not only paying for the charter expense but also subsidizing the administrative costs of the commission itself.

Appropriateness of Cooperative Research

Once a problem is identified and defined, the choice of whether the research project is conducted solely by scientists or cooperatively with fishermen depends on both the scientists and the fishermen. If the fishermen have a complaint about research that NMFS is already doing, they naturally want to show the government how to do the work more effectively. In the case of the surf clam fishery, the fishermen questioned the efficiency of the dredge used by NMFS to measure the surf clam recruitment and biomass. They wanted to use their own boats and design a more effective dredge for research. In this case, the fishermen's incentive to be part of a research project was to point out the shortcomings of existing fishery-independent research. This was also the case in several other examples, including New Zealand rock lobster, Canadian West Coast groundfish and East Coast halibut.

The North Pacific longliners likewise wanted to be part of research on seabird bycatch. In their case, the research was based on techniques of bird deterrence that already had been developed. Fishery-independent research would have been a redundant exercise. What the fishery needed were data on how well the techniques that they already had in the actual fishery worked. The volunteer vessels for the program also received an incentive. The time spent on the survey fulfilled the required quarterly observer coverage for each vessel, and the observers were paid by the research grant (instead of by each vessel).

The Pacific tuna seiners had no option. They either had to figure out how to fish without catching as many dolphins or live with large closed areas and an actively promoted boycott of their product. Since their problem related to fishing practices, the best way to solve the problem was to work on those practices while fishing and then effectively share, analyze, and utilize those solutions.

Long-term monitoring projects and ecological research are generally less well suited to cooperative projects, at least as they are currently funded. The budget stream is too unpredictable and such projects require commitments over long periods of time. There is not as much motivation or incentive for industry to participate because there is no immediate return. On the other hand, cooperative biological surveys and research on abundance and density of fish provide not only the immediate incentive to fishermen of money earned by working on a charter, but also the possibility of changes in management measures that may benefit both fishermen and the fishery resources on which they depend. A project is more likely to gain voluntary industry participation when there are appropriate financial incentives or expectations of later economic gains due to research findings.

Working Together

Once the scientists and fishermen have decided that cooperative research is the best course, they must decide how best to use each other's

---

BOX 8-1

Two summaries of "rules" for fishermen and scientists to successfully cooperate (from information provided to the committee):

1. Dr. Julia Parrish, University of Washington, in her explanation of the seabird surveys done in conjunction with Ed Melvin of Washington Sea Grant, with gillnetters in Puget Sound and with longliners in Alaska:

- Treat everyone with respect
- Make the research program valuable to the fishers
- Work in the active fishery
- Meet extensively with fishers before research starts, to brainstorm ideas
- Meet extensively with fishers during the program
- Meet with everyone else (agency personnel, environmentalists)
- Insist on scientifically rigorous experimental standards
- Follow through to implementation

2. Martin Hall, Inter-American Tropical Tuna Commission, in his explanation of working with tuna seiners to solve the problem of dolphin bycatch:

- Reasonable goals
- Practical approaches
- Respect
- Gradual improvement
- Participation
- Communication

---

expertise to best advantage. The fishermen are the experts on the fishing grounds; they understand the realities of working at sea. The scientists are the experts on experiment design and data gathering; they understand good scientific technique. Blending these two different points of view requires open-mindedness and tolerance, with trust and respect for everyone involved. Both the fishermen and scientists have to be willing not only to communicate with someone from outside their normal experience, they also have to provide and to accept diverse input and points of view. Both scientists and fishermen can benefit from this cross-fertilization of their respective ideas. In order for this kind of brainstorming to succeed, all parties must acknowledge each other's value. Ground rules of basic courtesy allow all participants to think and to speak freely, to take best advantage of everyone's expertise (Box 8-1).

Though the fishermen may give insight to a scientist about the design of an experiment, and a scientist may provide a perspective a fisherman has never considered, all participants in a cooperative research project also need to understand the limitations of their roles. Fishermen will be running their operations as a well-controlled scientific project. Scientists will be running an experiment on or collecting data from a real fishing platform. The fishermen must understand the parameters of the scientific work and the scientists must accept the capabilities and limitations of the fishing operations. Each side's input is valuable to the other, but the roles and limitations of each participant need to be clear from the beginning.

The model of NMFS scientists taking the lead in design and fishermen the lead in execution is far from universal or mandatory. NMFS is not the only science provider in the United States and in many cases university scientists have been heavily involved in the scientific design of cooperative research projects. In New Zealand and western Canada, numerous projects have been 100 percent industry designed and run relying on scientists from universities, the private sector, or in several cases in New Zealand government laboratories in Australia.

Project Management

Successful cooperative research projects often use a project leader who acts as a coordinator of the many roles involved in cooperative research. Initially, someone can help industry articulate the questions they have about research design and funding sources. Often, the time horizons of funding, management needs, data requirements, and priority setting are not aligned. Often, one knowledgeable individual needs to know ways to set the industry process in motion prior to the arrival of research funding. Through early planning, this individual could also improve opportunities for industry participation in priority setting.

A project leader can keep the goals of the cooperative research team clear and its purpose focused and defined. This person can also be a peacemaker and conciliator among divergent points of view in the design and implementation of the project. Mainly though, since communication throughout the project is essential to the project's success, someone who feels comfortable communicating with both fishermen and scientists is the most effective link to both sides of a cooperative research project. From the inception of the project and its initial planning stages to the setting of the techniques to be used in the research, the execution of the project and the deployment of boats and scientists, the assimilation of data at the conclusion of the project, and the analysis and distribution of those data, all parties involved need to communicate clearly and continually. One coordinator who thoroughly understands the importance of proper scientific work can be the hub of that communication.

This communicator can also be someone who maintains realistic expectations among the participants and the overseeing agency. This person can select the best and most practical ideas from all the participants of the project, including the fisheries managers, the scientists, and the fisherman, and assess which ones can be accomplished. Perhaps the scientists want more replications of a certain type of data than fishermen can accomplish in a day. Perhaps the fishermen are expecting more certainty in the results than the scientists believe they can achieve. Though responsibility should reside in one individual as the lead principal investigator, the most successful cooperative research results from an integrated team effort. The genius of cooperative research is that it attempts to harness divergent viewpoints to a common goal. Attention should be paid to mutual learning among team members and to defining roles based on expertise. More complex cooperative research projects often employ a small multidisciplinary board of directors or advisory group. Principal investigators in these situations should see themselves as a team leader and be generous when sharing credit.

As with any research, the quality of the work is only as good as the science employed. Successful cooperative research employs scientific standards as rigorous as those expected of any research. The scientists working with the fishermen need to be as scrupulous in their work as if they were doing research independently. From the design of the project to its implementation, the research needs to follow accepted scientific methods and standards. The scientists involved must design and execute research that will stand the scrutiny of peer review, just as any other research should.

Successful cooperative research requires fishermen who are professional operators and who run their businesses in a professional manner. They have to understand that their operations may have strict operating protocols. These restrictions can require them to operate differently but still work efficiently. Fishermen must be able to fish in a practical manner within the parameters of the research project. They also must be accountable to contracts and other obligations, must provide a safe working environment, and must follow strict guidelines while effectively using their fishing gear. Most important, the vessels used in successful cooperative research must be safe and seaworthy. Just as vessels used for fishery-independent research must follow the highest standards of safety, so must vessels used for cooperative research.

For cooperative research projects to be successful, scientists and the agency (or agencies) administrators must make a commitment to cooperative research. Cooperative research requires secured funding that is available from the planning to the completion of the research. Even in extreme cases where industry designs and operates the project, the management agencies must have a mechanism for accepting, evaluating, and incorporating the appropriate information provided.

Agencies that run successful cooperative research projects should have "cooperative research-friendly" policies. The personalities of scientists doing cooperative research are different from those of scientists who work exclusively in a lab or a classroom, and the overseeing agency needs to acknowledge and support this kind of scientist in order for cooperative research to have the greatest opportunity to succeed. Cases of scientists in need of material or information and in extenuating circumstances resulting from unforeseen circumstances at sea illustrate the need for the employers of scientists to provide cooperation and support in order to keep the project intact. The agency must be as committed as the scientist and likewise must be as flexible. Administrative infrastructure must have the capacity to handle cooperative partnerships.

Leadership support of scientists and managers who foster cooperative research is critical. This includes rewarding cooperative research with professional advancement and acknowledgment of the increased effort it takes to succeed in joint endeavors.

Consistent, basic standards and criteria for awarding grants, distributing money, selecting cooperative partners, choosing vessels, and deciding other aspects of cooperative projects, such as the basic authority to conduct such work, would shield the agency from challenges that the various cooperative research programs are inconsistent and therefore unfair. Further, communication among NMFS headquarters, regional offices, and the fisheries science centers should include sharing of problem-solving tactics and successful experiences. One region of the country need not be "reinventing the wheel" that another part of the country has already developed.

Successful cooperative research programs often require clear, complete contracts, written from the inception of the program, that define in complete detail the expectations from all parties as well as the compensation the fishermen will receive. These contracts spell out all the details of everyone's obligations and duties. This kind of contract ensures compliance by the fishermen with the project's protocols, prevents disputes about roles and duties, and provides assurances for the fishermen about their compensation.

The agency in charge of successful cooperative research also must be realistic about fishermen's compensation. They must be accountable and timely in their payment. A program that does not pay fishermen in a timely fashion will quickly lose the support of those fishermen. The fishermen who participate must also understand that the pay for their work is not renegotiable once the contract is signed.

Successful cooperative research requires that scientists, fishermen, and the overseeing agency be responsible and accountable in ways not normally required in order to maintain the focus and to address the original purpose of the research. In the execution of cooperative research, scientists must ensure that the experimental procedures are followed, despite any pressure to do otherwise. Fishermen cannot "improvise" on or vary from the experimental procedures in order to make their operation run more smoothly. They must follow through by executing on the fishing grounds the commitments they have made for the project. As with any research, there may need to be some adjustments to the experimental procedures to address design flaws or information needs. These changes must be agreed to by all participants. Further, the overseeing agency must utilize the results of the cooperative research appropriately. If a management agency has requested the results, it needs to follow through by assessing those results and using them appropriately.

Communication of Results

Proper and timely dissemination of the results from cooperative research projects is essential. The first step of this dissemination process is to have the results peer reviewed to validate the work. It is important that before any data or analysis reach anyone outside the project it first be fully analyzed and validated, through the peer review process and then put into final form. Once the review is completed, the results can be distributed to fishery managers, fishermen, and the public at large. If the cooperative research is based on a management need, it is especially important that the results be fully validated before managers receive them. There also may be pressure to release preliminary data or results prior to the completion of the peer review process. While this should be prevented, all participants need to be prepared to address possible interpretations of the leaked data and results and emphasize the importance of waiting until completion of the project (including peer review) prior to using the data and results.

Since an important benefit of cooperative research can be the fostering of healthy relations between NMFS and the fishing community, it is also important that the fishermen have full access to the results of the cooperative research through full distribution of the results to fishermen's associations and the press.

If the project is to be effective, the public and fishermen, especially those fishermen involved in the project, need to know that good research does not necessarily produce good news. Fishermen need to be able to do their part of the research without prejudice and must be willing to live with the results.

REASONS FOR FAILURE

Cooperative research can fail for the same reasons that any research can fail. If the science is not done well or if the data are poor, neither will be of much value. If the purpose of a research project is unclear or too broad, the results can be meaningless. If a project is not adequately funded, it may not be completed. These vulnerabilities are not unique to cooperative research. They apply to fishery-independent research as well. But the special conditions of cooperative research make it vulnerable to particular kinds of failure.

Lack of coordinating control is one way that a cooperative research project can fail. With the many different parties involved in cooperative research, any participant wavering from protocols can provide bad data. Cooperative research "can be frustrated or even destroyed by the actions of only a few individuals" (Leeman, 1998). If quality data are not collected all the way through the project, or at least consistently, the project is vulnerable to failure. The proper procedures for data gathering must be established early, communicated fully and repeatedly, and overseen properly, or the project can fail.

Another failure of control of cooperative research is when fishing done as research "assume[s] an economic life of its own . . . [when an] experimental fishery [is] regarded as a necessary part of the . . . industry" (Leeman, 1998).

In other words, the industry quickly becomes used to another source of income and resists, through fishing and through the political process, relinquishing that income. This can be prevented by a clear, complete statement of purpose to which all parties agree at the outset of the cooperative research. When the project ends, the industry has to be prepared to stop any extra fishing that the project may have provided. Once again, if this point is made clear early and communicated regularly, the problem can be avoided.

The results of research are often not simple. This is especially true if the cooperative research deals with a fishery with any allocation fights between user groups. Conclusions of a cooperative research project need to be carefully elucidated in order to draw accurate and proper conclusions. If unprocessed results of a project reach the public before the final reviews on that project are complete and validated, that project's intent is in jeopardy. What was intended to bring order to a problem can instead provoke chaos.

Fisheries with allocation wars between user groups make cooperative research difficult. Without careful screening of suggestions for cooperative research, there is a risk of doing research that may further the bargaining position of one specific user group. A group of fishermen may want to do research as a political attack on another group that targets the same stock of fish, with the result that the first group attains some allocation priority to those fish, or fishermen may oppose a specific proposal for research on their gear if they fear the data obtained may be used against them in allocation battles. For example, the results of the West Coast trawl bycatch study have often been used by users of other gear types as arguments against trawl allocation, leading some trawlers to question the usefulness of such studies for their political position in allocation battles. Allocation battles can ruin a project by degrading the commitment to good science and the proper gathering of data.

Cooperative research will effectively fail, no matter how good the scientific work is, if the administration required to handle the results of the work is insufficient. For fiscal years 1999 through 2003, the New England region has spent or obligated $7.9 million for cooperative research projects. With even more appropriations due in the future, regional officials have stated that they currently are unprepared to handle the amount of information from the resultant cooperative research. Results of successful projects will be useless if the management agencies cannot assimilate them as required. Indiscriminately providing money for cooperative research will not work if money and support are not provided appropriately to all parts of the system.

MAKING COOPERATIVE RESEARCH WORK

The extra education and administration needed for cooperative research will require extra work and vigilance on the part of NMFS, as has been noted earlier in this report. But a successful NMFS cooperative research program can provide solutions to problems that directly affect the agency.

The first of these problems is money. Budgets are shrinking, while the duties of NMFS are expanding and becoming more diverse and demanding. Conservation imperatives and management mandates are not as simple as they once were. Increasingly, rigorous standards of overfishing, bycatch, and ecosystem management require better management tools to meet their requirements. NMFS needs more research in order to develop these management tools, and in some cases cooperative research is the most cost-effective kind of research.

NMFS has to spend enormous amounts of manpower and money on recent litigation brought by nongovernmental organizations (NGOs) and fishermen. Cooperative research has effectively thwarted lawsuits when it has been a tool to target a specific contentious issue and has established a basis for making rational, effective decisions that can resolve a particular issue (e.g., albatrosses in Alaska, stock assessments in surf clams). The effect of the cooperative research is more than just providing good data. The process also can relieve the tension between the parties involved in the litigation by the act of working toward a common solution.

If NMFS can use cooperative research effectively for these purposes, it follows that the fishery management council (FMC) process will also benefit. By shedding light on contentious fisheries issues, cooperative research can provide a lubricant to the politics of the fisheries management process. If the parties involved in any FMC battle have worked together to understand an issue, they often have a less hostile, more understanding view of each other's position. Not only can cooperative research provide quality information with which an FMC can make decisions, but these decisions also can be made in a more conciliatory, less politically charged environment.

The FMCs were designed to be an open process, with direct participation of fishermen, scientists, and the public in the crafting of regulations. In practice, however, this sometimes is not the case if only because of a lack of understanding. Fishermen often need to know more of the management process and scientific principles on which management decisions are based. Many managers and scientists have not had direct experience in commer-

cial fisheries and often lack basic knowledge of fisheries under their review. Cooperative research provides a venue for helping to solve this problem.

Cooperative research is often cited as a means for fishermen to trust NMFS science, or at least to trust the results of research with which they are involved. When one considers the costs of a lack of trust—the costs in manpower and resources of a politically besieged management system—then effort expended to bolster science for the sake of building political will is worth doing. Trust has real value.

Cooperative research increases the knowledge of fishing communities of the processes that govern their fortunes. Well-informed fishing communities tend to be supportive of management and conservation, while fishing communities that are poorly informed tend to be reactionary and resistant.

Though most cooperative research has been done without the participation of NGOs, they could be enlisted as participants and provide another point of view for the design of cooperative research. Since many of the problems of fisheries stem from environmental concerns, the participation of NGOs could provide an opportunity to directly address those concerns. The resulting partnerships could be another way to defuse potentially contentious and litigious issues. If the NGOs have the same understanding of a problem as the fishermen and the scientists, NMFS will have a better ability to reach a consensus with all these groups without going to court.

The indiscriminate use of cooperative research poses some dangers. "Science for hire" with a preconceived agenda could pass for cooperative research if the screening and review process is not sufficiently rigorous. Though legitimate cooperative research can have the benefit of uniting people, misused or shoddy cooperative research could further divide people, cause more hostility, and be an unsound basis for management.

Cooperative research possesses potential problems in other ways. Results of cooperative research may prove that a scientist with a long history of a certain point of view has been mistaken. Scientists must be willing to accept that result if the cooperative research is valid. Fishermen may find that the results of cooperative research show that a stock of fish may be weaker or a certain gear may be more destructive than they thought. They may face restrictions they don't like as a result of cooperative research. As with any research, if the work is done well, the results must be taken for what they are. If fishermen realize this and are favored by the status quo, they may resist cooperative research. If scientists realize this and are too proud to undergo scrutiny in the open forum of cooperative research, they may resist cooperative research as well.

9

Findings and Recommendations

Below are the findings and recommendations developed by the committee based on information provided in this report. The findings and recommendations are grouped in the following four categories: Evaluating the Benefits of Cooperative Research, Allocation of Funding, Legal and Administrative Issues, and Communication. Each finding is followed by one or more associated recommendations.

EVALUATING THE BENEFITS OF COOPERATIVE RESEARCH

Finding: Most fisheries research projects can benefit from some level of cooperation.

Consistent with assumptions and stated goals in the agency's strategic plans, the National Marine Fisheries Service (NMFS) should explicitly recognize that fishermen and other stakeholder groups can be engaged in many types of fisheries research. The degree of cooperation will depend on regional and fisheries-specific needs and opportunities and the potential gains in achieving science and management objectives. In particular, fishermen generally have extensive knowledge of fishing gear, fishing grounds, and fish behavior, and this knowledge can be incorporated in most forms of research.

Recommendation: Cooperative research should be considered as a usual and normal approach for conducting fisheries research.

Cooperative research has many advantages and the proportion of fisheries research projects that will be conducted cooperatively should be increased. However, this research should not be done at the expense of required management mandates and core research.

Finding: Some cooperative research projects may have small effects on improving the science but may substantially achieve other fishery management objective(s) (e.g., improving stakeholder trust in the fishery science and management system; improving research methods and administration; coeducating scientists, industry, and managers).

Many types of research will have complementary effects in achieving science and management objectives; in other cases there may be significant tradeoffs.

Recommendation: In designing cooperative research projects, the applicability of the results to the overall success of the fishery must be considered.

The benefits of cooperative research should not be based solely on their impact on improving science, but rather their impact on improving management goals and objectives.

ALLOCATION OF FUNDING

Finding: Congressional appropriations through earmarks and line items in the NMFS budget have supported cooperative research but have the following drawbacks: they are inconsistent with the research needs across regions, are not predictable from year to year, may not provide a fair and equitable process for allocation of research funding, may possibly deduct from the NMFS base budget, and may not account for agency costs of supporting cooperative research projects.

Recommendation: Congress and NMFS should give serious consideration to establishing and funding regional research boards to: prioritize and coordinate the use of dedicated funding (earmarks and line items) for cooperative research projects in each region, evaluate NMFS-dedicated research projects for their potential as cooperative research, foster communication of research results, and evaluate coop-

erative research projects and programs. A national steering committee consisting of the chairs of each regional research board should also be formed to provide coordination among regions and facilitate communication with the NMFS national office.

Fishery research needs vary from region to region. The research that is funded should reflect the needs of NMFS, the regional constituents, and the resource. Currently, federal funding for cooperative research is provided through a variety of organizations, including NMFS. Coordination is needed to make sure that the sum of the research efforts from federal funding addresses the most pressing research needed to support fishery management objectives and that duplication of effort is avoided. Regional research boards could serve these functions.

The regional research boards should have provisions allowing them to receive private funding as well and to support multiyear projects. The regional research boards could be administered independently or through the regional fishery management councils.

Finding: Funds for fisheries research are limited, and cooperative research funding needs to be used in ways that significantly improve the fisheries management process.

Cooperative research projects need to be designed to address important management issues and must meet high scientific standards, or the results may not be considered valid and will not affect the management system. If cooperative research projects are poorly designed, the costs may be much greater than necessary and may lead to failure of the project.

Recommendation: A majority of cooperative research funds should be allocated through a competitive review process. The remaining monies should be used for rapid response, seed money, and administration.

Standards of practicality, cost effectiveness, utility in the management process, and scientific rigor must be considered when allocating cooperative research funds. Rigorous peer review of cooperative research proposals will help ensure that funded projects will have the highest likelihood of achieving project and research objectives. Data and analysis must also undergo peer review. As demands for peer review on all elements of the fishery management system grow, cooperative research must be subject to the same criteria.

Adequate administrative funding needs to be provided to facilitate cooperative research projects and programs on a timely basis. Some funding needs to be made available for rapid response and to support pilot studies.

LEGAL AND ADMINISTRATIVE ISSUES

Finding: In the course of experience with cooperative research in the United States, numerous administrative, legal, and permitting issues have arisen that have impeded these projects.

For cooperative research to achieve its full potential, changes need to be made in the administrative and legal structure. The following four recommendations address specific administrative and legal structure needs:

Recommendation: Commercial fishing vessels used for cooperative fisheries research by NMFS should meet all U.S. Coast Guard requirements for operation and manning so as to ensure safe operations.

All participants in cooperative research need to be provided with a safe working environment.

Recommendation: NMFS should ensure that appropriate liability insurance is secured for all cooperative fisheries research activities so as to protect the financial interests of all participants involved in cooperative research.

While the sources of insurance may vary, NMFS needs to ensure that all participants in cooperative research possess liability insurance applicable to the research setting prior to the onset of the research project.

Recommendation: NMFS should streamline and standardize all permitting procedures for conducting cooperative research projects so as to ensure uniform treatment and rapid processing of all applications in all regions.

Delays in permitting can result in increased costs to participants, delay payments to fishermen and other participants, and result in some research projects being cancelled. The latter could occur if, for example, the research project required the presence of a certain species of migratory fish. If a permit is delayed too long, the fish could migrate away from the area before the research could start. The fishermen involved in the cooperative research project may have forgone other fishing opportunities (and resulting income) so that they could participate in the research and may have been depending on the income from the project. There also needs to be consistency in permitting procedures across regions.

Recommendation: NMFS and operators of commercial fishing vessels should use comprehensive contracting procedures so as to minimize confusion and maximize opportunity for all fishermen to participate in cooperative research.

Finding: Observers are used to collect data in a wide variety of cooperative research projects, and the role of observers in fisheries management is increasing. It is particularly important that the observer data be accepted as valid by all parties involved in the fishery.

Recommendation: All participants in cooperative research should ensure the independent status of observers and the confidentiality of the data collected.

Finding: Large cooperative research projects involve a range of parties, including the fishing industry, government scientists, regulators, university scientists, and other stakeholders.
The successful completion of large projects requires the resources and assistance of many parties and seeks to have the results accepted by all parties. Successful coordination of complex projects is difficult but essential.

Recommendation: For larger and more complicated cooperative research programs, specific advisory committees should be formed.
The advisory committees should have representatives from all participating parties, and members should be selected based on their ability to bring resources to the project, their ability to facilitate the project (perhaps through permitting), and their potential role as communicators of project results.

Finding: Unsuccessful cooperative research projects have often failed to meet expectations because of misunderstandings about individual or group responsibilities.
Cooperative research involves a number of parties and thus more complex organization than single-agency projects.

Recommendation: Expectations, requirements, and procedures, including the development of agreements carefully detailing the responsibilities of all participants, should be clarified at the beginning of every project.

These agreements and contracts should include project design, project implementation, contingency plans, data collection, analysis, and communication responsibilities and decision-making authority.

Finding: Cooperative research will effectively fail, no matter how good the scientific work is, if the administration required to handle the work is insufficient.

Adequate administrative support needs to be provided at all phases of cooperative research. Insufficient administrative support can result in delays in the permitting, funding decisions, and payments.

Recommendation: Appropriate administrative overhead should be included in all budgetary allocations.

Finding: Institutional support of scientists and managers who foster cooperative research is critical.

NMFS scientists who participate in cooperative research need to know that their involvement is supported by their supervisors and that their professional advancement will be commensurate with that of NMFS scientists who do not participate in cooperative research.

Recommendation: NMFS scientists who participate in cooperative research should be given equal opportunity for professional advancement along with their counterparts who do not participate in cooperative research.

Cooperative research requires additional time and effort to be successful, and there may be fewer opportunities to publish in substantial peer-reviewed journals. This needs to be recognized by NMFS and considered when evaluating these scientists for professional advancement. NMFS should encourage student involvement in cooperative research in order to develop the necessary human capital to lead future cooperative research.

COMMUNICATION

Finding: Communication is critical at all stages of cooperative research.

Opportunities for cooperative research and the availability of project funding need to be communicated; NMFS needs to respond to industry suggestions; participants need to communicate regularly during the design and implementation of a project; and the results need to be communicated and disseminated upon completion of the research. This finding resulted in the following two recommendations:

Recommendation: NMFS should recognize and hire individuals with the interpersonal and communication skills necessary for cooperative research.

In order to assuage possible communication problems, NMFS either needs to invest in developing these capacities in-house or it needs to build effective relationships with entities that do. These entities need to have an understanding of fisheries science and management.

Recommendation: NMFS should require that a communications plan for outreach, progress reports, and dissemination of the final results be part of every cooperative research project plan.

The final success of cooperative research projects relies on the proper dissemination and utilization of its results.

References

Bergh, M.O., E.K. Pikitch, J.R. Skalski, and J.R. Wallace. 1990. The statistical design of comparative fishing experiments. *Fisheries Research* 9(2):143-163.

Bernstein, B., and S. Iudicello. 2000. National Evaluation of Cooperative Data Gathering Efforts in Fisheries: A Report to the National Marine Fisheries Service. National Fisheries Conservation Center Project Report. Unpublished.

Fisher, R.B., J. Harms, M. Hosie, B. McCain, R. Schoning, and G. Sylvia (eds.). 1999. Working Together for West Coast Groundfish: Developing Solutions to Research Needs in 1998, 1999, 2000, and Beyond. Conference Proceedings, Portland, Oregon, July 16-17, 1998. Newport, OR: Hatfield Marine Science Center.

Hall, M.A. 1998. An ecological view of the tuna-dolphin problem: Impacts and trade-offs. *Review of Fish Biology and Fisheries* 8:1-34.

Harms, J., and G. Sylvia. 1999. Industry-Scientist Cooperative Research: Application to the West Coast Groundfish Fishery. Report to NOAA Fisheries Northwest Fisheries Science Center. Newport, OR: Cooperative Institute for Marine Resource Studies.

Harte, M. 2001. Collaborative Research: Innovations and Challenges for Fisheries Management in New Zealand. Microbehavior and Macroresults: Proceedings of the Tenth Biennial Conference of the International Institute of Fisheries Economics and Trade, July 10-14, Corvallis, OR.

Leaman, B.M. 1998. Experimental rockfish management and implications for rockfish harvest refugia. P. 23 in NOAA Technical Memorandum: Marine Harvest Refugia for West Coast Rockfish: A Workshop, Aug. 1998, M. Yoklavich (ed.). Silver Spring, MD: NOAA, National Marine Fisheries Service. 159 pp.

Melvin, E.F., J.K Parrish, K.S. Dietrich, and O.S. Hamel. 2001. Solutions to seabird bycatch in Alaska's demersal longline fisheries. Washington Sea Grant Program. Project A/FP-7.

National Academy of Public Administration (NAPA). 2002. Courts, Congress, and Constituencies: Managing Fisheries by Default. National Academy of Public Administration, Washington, D.C.

National Marine Fisheries Service (NMFS). 2001. NMFS Strategic Plan for Fisheries Research. U.S. Department of Commerce. Silver Spring, MD: NOAA, National Marine Fisheries Service.

National Oceanic and Atmospheric Administration (NOAA). 2002. Fiscal Year 2003 Budget Summary. U.S. Department of Commerce. Silver Spring, MD. Available at http://www.publicaffairs.noaa.gov/budget2003/noaabluebook.pdf.

National Research Council (NRC). 2002. Science and Its Role in the National Marine Fisheries Service. Washington, D.C.: National Academy Press.

NOAA Fisheries. 2002. NOAA Fisheries Strategic Plan. Silver Spring, MD: NOAA Fisheries.

Northwest Fisheries Science Center (NWFSC). 2000. West Coast Groundfish Research Plan. Seattle, WA: NOAA Fisheries Northwest Fisheries Science Center.

Pikitch, E.K. 1986. Impacts of management regulations on the catch and utilization of rockfish in Oregon. Pp. 369-387 in Proceedings of the International Rockfish Symposium, October 20-22, 1986, Anchorage, Alaska.

Pikitch, E.K. 1991. Technological interactions in the U.S. West Coast groundfish fishery and their implications for management. ICES Marine Science Symposium 193:253-263.

Pikitch, E.K. 1992. Potential for gear solutions to bycatch problems. Pp. 128-138 in Proceedings of the National Industry Bycatch Workshop, February 4-6, 1992, Schoning et al. (ed.), Newport, OR.

Pikitch, E.K, D.L. Erickson, and J.R. Wallace. 1988. An evaluation of the effectiveness of trip limits as a management tool. P. 33 in NWAFC Processed Rep. 88-27. Seattle, WA: Northwest and Alaska Fisheries Center, National Marine Fisheries Service.

Pikitch, E.K., M.O. Bergh, D.L. Erickson, J.R. Wallace, and J.R. Skalski. 1990. Final report on the results of the 1988 West Coast groundfish mesh size study. Technical Report FRI-UW-9019, WSG-MR 90-02.

Saelens, M. 1995. Pilot Project Proposal: West Coast Volunteer Logbook Program. Newport, OR: Marine Division, Oregon Department of Fish and Wildlife.

Turris, B.R. 1999. A Comparison of British Columbia's ITQ Fisheries for Groundfish Trawl and Sablefish: Similar Results from Programs with Differing Objectives, Designs, and Processes. Proceedings of the Fish Rights 99 Conference, November 14-17, Freemantle Western Australia.

Wondolleck, J.M., and S.L. Yaffee. 2000. Making Collaboration Work: Lessons from Innovation in Natural Resource Management. Washington, D.C.: Island Press. 277 pp.

Zwanenburg, K.C.T., and S. Wilson. 2000. The Scotian Shelf and Southern Grand Banks Halibut (Hippoglossus hippoglossus) survey—collaboration between the fishing and fisheries science communities. International Council for Exploration of the Sea CM 2000 / W:20, 26 pp.

Appendixes

A

Statement of Task

This study will address issues essential for the effective design and implementation of cooperative and collaborative research programs. Cooperative research programs are currently being administered by the National Marine Fisheries Service (NMFS) to foster the participation of fishermen in the collection of scientific information used in fisheries management. The committee's report will identify design elements necessary for achieving programmatic goals with scientific validity, including (1) identification of data needs; (2) setting of research priorities; (3) identification of potential participants; (4) matching of research needs with fishing expertise and access to appropriate vessel and gear; and (5) methods for communicating the findings from this research to the communities involved in or affected by fisheries management.

In addition, the report will address issues essential for the effective implementation of the program such as (1) the maintenance of scientific validity; (2) criteria for awarding and distributing research funds; (3) procedures for evaluating, ranking, and funding research proposals; and (4) ownership and confidentiality of data collected with funds from the cooperative research program. As part of this study, the committee will examine the implementation of existing cooperative research programs.

B

Project Oversight

OCEAN STUDIES BOARD

NANCY RABALAIS (*Chair*), Louisiana Universities Marine Consortium, Chauvin

ARTHUR BAGGEROER, Massachusetts Institute of Technology, Cambridge

JAMES COLEMAN, Louisiana State University, Baton Rouge

LARRY CROWDER, Duke University, Beaufort, North Carolina

RICHARD B. DERISO, Inter-American Tropical Tuna Commission, La Jolla, California

ROBERT B. DITTON, Texas A&M University, College Station

EARL DOYLE, Shell Oil (ret.), Sugar Land, Texas

ROBERT DUCE, Texas A&M University, College Station

PAUL G. GAFFNEY II, National Defense Univeristy, Washington, DC

WAYNE R. GEYER, Woods Hole Oceanographic Institution, Massachusetts

STANLEY R. HART, Woods Hole Oceanographic Institution, Massachusetts

MIRIAM KASTNER, Scripps Institution of Oceanography, La Jolla, California

RALPH S. LEWIS, Connecticut Geological Survey, Hartford

WILLIAM F. MARCUSON III, U.S. Army Corps of Engineers (ret.), Vicksburg, Mississippi

JULIAN P. MCCREARY, JR., University of Hawaii, Honolulu
JACQUELINE MICHEL, Research Planning, Inc., Columbia, South
   Carolina
SCOTT NIXON, University of Rhode Island, Narragansett
SHIRLEY A. POMPONI, Harbor Branch Oceanographic Institution,
   Fort Pierce, Florida
FRED N. SPIESS, Scripps Institution of Oceanography, La Jolla,
   California
JON G. SUTINEN, University of Rhode Island, Kingston
NANCY TARGETT, University of Delaware, Lewes

Staff

MORGAN GOPNIK, Director
JENNIFER MERRILL, Senior Program Officer
SUSAN ROBERTS, Senior Program Officer
DAN WALKER, Senior Program Officer
JOANNE BINTZ, Program Officer
TERRY SCHAEFER, Program Officer
ROBIN MORRIS, Financial Officer
JOHN DANDELSKI, Research Associate
SHIREL SMITH, Administrative Associate
NANCY CAPUTO, Senior Project Assistant
DENISE GREENE, Senior Project Assistant
BYRON MASON, Senior Project Assistant
SARAH CAPOTE, Project Assistant
TERESIA WILMORE, Project Assistant

C

Committee and Staff Biographies

COMMITTEE

Ray Hilborn (*Chair*) is currently Richard C. and Lois M. Professor of Fisheries Management at the School of Aquatic Fishery Sciences of the University of Washington. Dr. Hilborn earned his Ph.D. in zoology from the University of British Columbia in 1974. His main areas of research are resource management, population dynamics, and conservation biology. Dr. Hilborn was a member of the Committee on Fish Stock Assessment Methods and the Ocean Studies Board.

Joseph DeAlteris has been a professor at the University of Rhode Island since 1995. Dr. DeAlteris earned his Ph.D. in 1986 from the Virginia Institute of Marine Science. His recent research has focused on aquatic resource harvesting technologies and their impact on the ecosystem, in particular the reduction of bycatch through development of size- and species-specific fishing gear and the quantitative evaluation of effects of fishing gear on fish stocks, habitat, and manmade structures placed on and under the seabed. Dr. DeAlteris was a member of the committee for the report *Effects of Trawling and Dredging on Seafloor Habitat.*

Richard Deriso is chief scientist of the Tuna-Billfish Program of the Inter-American Tropical Tuna Commission. He also serves as adjunct associate professor at the Scripps Institution of Oceanography and affiliate professor

127

of fisheries at the University of Washington. Dr. Deriso earned his Ph.D. in biomathematics from the University of Washington in 1978. His major research interests are in the areas of fisheries population dynamics, quantitative ecology, stock assessment, applied mathematics, and statistics. Dr. Deriso is a member of the Ocean Studies Board and served on the Committee to Review Atlantic Bluefin Tuna.

Gary Graham is a professor and extension marine fisheries specialist at Texas A&M University, Galveston. Mr. Graham earned his B.S. in range management at Texas A&M University. His research interests are fisheries development activities focusing on assessment of resources, determination of harvesting gear, and establishment of markets. Mr. Graham was a member of the Committee on Sea Turtle Conservation.

Suzanne Iudicello is currently an independent consultant after serving for 10 years with the Center for Marine Conservation, until 1997 as its vice president for programs and general counsel. Ms. Iudicello earned her J.D. in environmental law from George Washington University in 1989. She organized and conducted several successful negotiations between fishers and conservationists in 1988 and 1993, resulting in amendments to the Marine Mammal Protection Act that have reduced incidental takes of marine mammals in fishing operations.

Mark Lundsten actively fished for 27 seasons, mostly for halibut and sablefish in the Gulf of Alaska and the Bering Sea. He recently retired as the owner-operator of the fishing vessel *Masonic*, a 70-foot longliner. Mr. Lundsten earned his B.A. in English from Pomona College in 1973. He helped develop bird deterrence techniques in the longline fishery and then ran his vessel as a survey boat to test and refine those techniques, which later became regulations. He is a board member of the National Fisheries Conservation Center and served on the National Marine Fisheries Service IFQ Panel to review the National Research Council report *Sharing the Fish: Toward a National Policy on Individual Fishing Quotas*.

Ellen Pikitch is director of marine conservation programs at the Wildlife Conservation Society and in that capacity oversees the society's field and laboratory marine and freshwater research and conservation efforts. Dr. Pikitch earned her Ph.D. in zoology from Indiana University in 1982. Dr. Pikitch's main research interests are in fisheries science, stock assessment, bycatch problems, and other marine conservation issues. Dr. Pikitch served on the Committee on Ecosystem Management for Sustainable Marine Fisheries.

Gil Sylvia is the superintendent of the Coastal Oregon Marine Experiment Station and associate professor in agricultural resources and economics at Oregon State University. Dr. Sylvia earned his Ph.D. in marine resource economics from the University of Rhode Island in 1989. His research and outreach have focused on a broad range of fisheries and aquaculture issues, with particular emphasis on market development and public policy for Oregon fisheries. Dr. Sylvia is a member of the Science and Statistical Committee of the Pacific Fisheries Management Council.

Priscilla Weeks has worked as a research associate at the Environmental Institute of Houston, University of Houston, Clear Lake, since 1993. Dr. Weeks earned her Ph.D. in anthropology in 1988 from Rice University. Her research focuses on cultural anthropology, social aspects of natural resource management and environmental regulations, social aspects of natural resource management and rural development, and cross-cultural scientific collaboration. She is also a member of the socioeconomic panel that provides advice to the Gulf of Mexico Fishery Management Council. Dr. Weeks served on the committee for the report *Effects of Trawling and Dredging on Seafloor Habitat.*

John Williamson is a member of the New England Fishery Management Council, serving on the Sea Scallop, Groundfish, Herring, Protected Species, Habitat, and Gear Conflict Committees and is chairman of the Research Steering Committee. Mr. Williamson's current interest is working with commercial fishermen to promote collaborative research and data-gathering programs. Mr. Williamson has been working under a series of grant awards administered through the New England Aquarium Conservation Department to promote conservation planning by the fishing industry.

Kees Zwanenburg is a research scientist in the Marine Fish Division at the Bedford Institute of Oceanography. Mr. Zwanenburg earned his B.Sc. in zoology (honors) from the University of Alberta in 1977 and his M.Sc. in fish population dynamics from McGill University in 1981. During the past 20 years, Mr. Zwanenburg has identified shortcomings in our understanding of exploited populations and conducted research to address them. He was among the first to recognize the value of collaborative research with the fishing industry and has implemented several major initiatives. His present research focuses on how fisheries impact marine ecosystems.

NATIONAL RESEARCH COUNCIL STAFF

Terry Schaefer is a program officer at the Ocean Studies Board where he has been since August 2001. He received his Ph.D. in oceanography and coastal sciences from Louisiana State University in 2001 and a master's degree in biology/coastal zone studies from the University of West Florida in 1996. In 1998 he served as a John A. Knauss Marine Policy Fellow in the office of the chief scientist, National Oceanic and Atmospheric Administration. Since joining the Ocean Studies Board, he has directed the study *Science and Its Role in the National Marine Fisheries Service* (2002). Previously, Dr. Schaefer worked for the U.S. Environmental Protection Agency, National Park Service, U.S. Army Corps of Engineers, U.S. Fish and Wildlife Service, and U.S. Forest Service. Dr. Schaefer's interests include recruitment dynamics of marine populations, experimental statistics, coastal zone management, and marine policy.

Denise Greene is a senior project assistant at the Ocean Studies Board and has nine years of experience working for the National Academies. Mrs. Greene has been involved with studies on marine biotechnology, environmental information for naval warfare, and fisheries policy.

D

Acronyms

CGCS   Canadian Groundfish Conservation Society
CMAST   Center for Marine Science and Technology
CMC   Center for Marine Conservation
CSA   Canadian Sablefish Association

DFO   Department of Fisheries and Oceans (Canada)
DTS   Dover sole, thornyheads, sablefish

EA   environmental assessment
EFP   exempted fishing permit

FMC   fishery management council
FSRS   Fishermen and Scientists Research Society

GT   gross tons

IATTC   Inter-American Tropical Tuna Commission
IFQ   individual fishing quota
ITQ   individual transferable quota
IPHC   International Pacific Halibut Commission

LOA   letter of acknowledgment

MFish	Ministry of Fisheries (New Zealand)
MMD	merchant mariner's documents
MSFCMA	Magnuson-Steven Fishery Conservation and Management Act

NAPA	National Academy of Public Administration
NEFSC	Northeast Fisheries Science Center
NGO	nongovernmental organization
NIWA	National Institute of Water and Atmospheric Research (New Zealand)
NMFS	National Marine Fisheries Service
NOAA	National Oceanic and Atmospheric Administration
NPFMC	North Pacific Fishery Management Council
NRC	National Research Council
NWFSC	Northwest Fisheries Science Center

OCMI	officer in charge, marine inspection
ODF&W	Oregon Department of Fish and Wildlife
OTC	Oregon Trawl Commission

| PFMC | Pacific Fisheries Management Council |

| RLIC | Rock Lobster Industry Council (New Zealand) |

| SeaFIC | Seafood Industry Council (New Zealand) |

| TAC | total allowable catch |
| TED | turtle excluder device |

| USFWS | U.S. Fish and Wildlife Service |